ROUTLEDGE LIBRARY EDITIONS: INTERNATIONAL SECURITY STUDIES

Volume 8

THE FUTURE OF AIR POWER

THE FUTURE OF AIR POWER

NEVILLE BROWN

Routledge
Taylor & Francis Group
LONDON AND NEW YORK

First published in 1986 by Croom Helm Ltd

This edition first published in 2021
by Routledge
2 Park Square, Milton Park, Abingdon, Oxon OX14 4RN

and by Routledge
52 Vanderbilt Avenue, New York, NY 10017

Routledge is an imprint of the Taylor & Francis Group, an informa business

© 1986 Neville Brown

All rights reserved. No part of this book may be reprinted or reproduced or utilised in any form or by any electronic, mechanical, or other means, now known or hereafter invented, including photocopying and recording, or in any information storage or retrieval system, without permission in writing from the publishers.

Trademark notice: Product or corporate names may be trademarks or registered trademarks, and are used only for identification and explanation without intent to infringe.

British Library Cataloguing in Publication Data
A catalogue record for this book is available from the British Library

ISBN: 978-0-367-68499-0 (Set)
ISBN: 978-1-00-316169-1 (Set) (ebk)
ISBN: 978-0-367-70957-0 (Volume 8) (hbk)
ISBN: 978-0-367-70961-7 (Volume 8) (pbk)
ISBN: 978-1-00-314870-8 (Volume 8) (ebk)

Publisher's Note
The publisher has gone to great lengths to ensure the quality of this reprint but points out that some imperfections in the original copies may be apparent.

Disclaimer
The publisher has made every effort to trace copyright holders and would welcome correspondence from those they have been unable to trace.

The Future of Air Power

Neville Brown

CROOM HELM
London & Sydney

© 1986 Neville Brown
Croom Helm Ltd, Provident House, Burrell Row,
Beckenham, Kent BR3 1AT

Croom Helm Australia Pty Ltd, Suite 4, 6th Floor,
64–76 Kippax Street, Surry Hills, NSW 2010, Australia

British Library Cataloguing in Publication Data

Brown, Neville
  The future of air power.
  1. Air power
  I. Title
  358.4   UG630
  ISBN 0-7099-3217-0

Phototypeset in English Times by Pat and Anne Murphy
10 Bracken Way, Highcliffe-on-Sea, Dorset
Printed and bound in Great Britain by Mackays of Chatham Ltd, Kent

CONTENTS

List of Abbreviations
The Author
Preface

Part One: The Background to Debate

1. *The Historical Setting* — 3
   The Ascent of Air Power — 3
   The Evolution of Analysis — 17
   Sortie and Loss Rates — 23
2. *Contemporary Air Doctrines* — 33
   The West — 33
   Soviet Perceptions — 41
3. *The Human Factor* — 49
   Sustained Offensive Action — 49
   The Citadel Image — 51
   Aircrew Stress — 52
   Training and Recruitment — 54
   One Seat or Two? — 55
   Adaptation to War — 57
4. *The Impact of Novel Technology* — 63
   Electronic Warfare — 63
   The Future of Aeronautical Science — 71
   Other Advances in Air Ordnance — 82
5. *The Changing Land Battle* — 94
   Movement and Fire — 94
   The Liddell Hart Fallacy — 95
   The Implications for Air Power — 98
   The Triumph of Artillery? — 100
   Geostrategic Perspectives — 102
6. *Technology Gaps* — 105
   The Global Sweep — 105
   The Central Comparison — 106
   Electronics Again — 109
   The Extended S-Curve — 111
   No Grounds for Euphoria — 113

Part Two: Pre-Nuclear War Over Land

7. *The Classic Missions* — 119
   - The Geographical Setting — 119
   - Surface versus Air — 123
   - Interdiction — 136
   - Tactical Reconnaissance — 142
   - Air Defence in Depth — 149
8. *Tactical Air Mobility* — 163
9. *The Vulnerability of Airfields* — 178

Part Three: Other Environments

10. *Air Power at Sea* — 193
11. *Aerial Mass Destruction* — 206
    - The Demise of Strategic Air Deterrence — 206
    - Theatre Confrontation — 215

Part Four: A Predictive Overview

12. *Costs versus Benefits* — 227
    - Investment and Operation — 227
    - Full-life Cost — 235
    - Economy through Multi-role? — 239
    - Simple is Effective? — 240
13. *A Multiple Revolution in War* — 247
    - Air Power Ascendant — 247
    - Warplanes in Decline? — 250
    - The Sudden Blitzkrieg — 255
    - NATO's Quest for Strategy — 263
    - Maritime Air Cover — 266
    - A Mix of Systems — 267
    - Air Power and Geography — 272

Appendix A: *The Electromagnetic Environment* — 279
Appendix B: *Terminology and Notation* — 291
Appendix C: *Some Contemporary Military Aircraft* — 294
   - Strategic Bombers — 294
   - Tactical Monoplanes — 295
   - Helicopters — 298
   - Airborne Early Warning — 299
   - Heavy Transport Aircraft — 300

Index — 301

LIST OF ABBREVIATIONS

ADV	Air Defence Version (of the Tornado)
AFV	Armoured Fighting Vehicles
AGARD	Advisory Group for Aeronautical Research and Development
ALCM	Air-Launched Cruise Missile
ASW	Anti-Submarine Warfare
ATB	Advanced Technology Bomber
AWACS	Airborne Warning and Control System
Å	Angström
CAP	Combat Air Patrol
CAS	Close Air Support
CCV	Control-Configured Vehicle
CEP	Circular Error Probability
CVA	Carrier Vessel Attack
CW	Continuous Wave
ECCM	Electronic Counter-Counter Measures
ECM	Electronic Counter Measures
EFA	European Fighter Aircraft
EHF	Extremely High Frequency
EMP	Electromagnetic Pulse
ET	Emerging Technologies
EW	Electronic Warfare
FAE	Fuel Air Explosive
FBM	Fleet Ballistic Missile
FEBA	Forward Edge of the Battle Area
FLOT	Forward Line of Own Troops
FLED	Forward Line of Enemy Defence
FLIR	Forward-Looking Infra-Red
FOFA	Follow-on Forces Attack
FPB	Fast Patrol Boat
GCA	Ground-Control Approach
GLCM	Ground-Launched Cruise Missile
HE	High Explosive
HF	High Frequency
ICBM	Intercontinental Ballistic Missile
IFF	Identification Friend or Foe

List of Abbreviations

IISS	International Institute for Strategic Studies
INEWS	Integrated Electronic Warfare System
JATO	Jet-Assisted Take-Off
JVX	Joint Services Advanced Vertical Light Aircraft
kc/s	kilocycles per second
kg	kilogram
km	kilometre
kmc/s	kilomegacycles per second
km/h	kilometres per hour
kPa	kilopascal
LRMP	Long-Range Maritime Patrol
m	metre
MASH	Mobile Army Surgical Hospital
MBT	Main Battle Tank
mc/s	megacycles per second
MITI	Ministry of International Trade and Industry
MLRS	Multiple Launch Rocket System
mm	millimetre
mph	miles per hour
mps	metres per second
MTBF	Mean Time Between Failures
NATO	North Atlantic Treaty Organisation
NOE	Nap of the Earth
OECD	Organisation for Economic Co-operation and Development
OMG	Operational Manoeuvre Group
OTH	Over-the-Horizon
OTH-B	Over-the Horizon-Backscatter
PGM	Precision Guided Munition
PLA	People's Liberation Army
PLO	Palestine Liberation Organisation
PVO-Strany	*Protivo Vozdushnaya Oborona Strany*
QRA	Quick Reaction Alert
RAF	Royal Air Force
REB	*Radio electronnaya bor'ba*
RPV	Remotely Piloted Vehicle
RSAF	Royal Swedish Air Force
RSN	Royal Swedish Navy
SAC	Strategic Air Command
SAM	Surface-to-Air Missile
SHAPE	Supreme Headquarters Allied Powers Europe

List of Abbreviations

SHF	Super High Frequency
SIAN	Self-Initiating Anti-Aircraft Missile
SLAR	Sideways-Looking Airborne Radar
SLBM	Submarine-Launched Ballistic Missile
SOC	Sector Operations Centre
SRAM	Short-Range Attack Missile
STOL	Short Take-Off and Landing
USA	United States of America
USAAF	United States Army Air Force
USAF	United States Air Force
USN	United States Navy
VDS	Variable Depth Sonar
VLF	Very Low Frequency
V/STOL	Vertical or Short Take-Off and Landing
VTOL	Vertical Take-Off and Landing

THE AUTHOR

Neville Brown has a personal chair of International Security Affairs at the University of Birmingham. He has held Visiting Fellowships at the Stockholm International Peace Research Institute, the Australian National University and the International Institute for Strategic Studies. Between 1972 and 1978, he was an Academic Consultant to the National Defence College.

He read Economics at University College, London, and History at New College, Oxford. From 1957 to 1960 he was a forecaster in the meteorological branch of the Fleet Air Arm; he has been a Fellow of the Royal Meteorological Society since 1960. Subsequently he was a staff member of the Royal Military Academy, Sandhurst, and then of the Institute for Strategic Studies, as it then was. He is currently Chairman of the Council for Arms Control.

From 1965 to 1972, he served as a defence correspondent in various parts of Afro-Asia for several leading Western journals. He is the author of a number of books and monographs on strategic studies and contemporary history. His next major study will be entitled *The Challenge of Climatic Change*.

PREFACE

The middle years of my childhood were encompassed only too neatly by World War II. To live then just under the scarp of the Oxfordshire Chilterns was to acquire a strong awareness of the majesty of air power as exercised more or less independently of the land and sea dimensions in warfare. Within a few miles of my native Watlington a key photo-reconnaissance and a major airborne base were established, as well as searchlight units and, so I now gather, a dummy airfield. Just to the south-east lay the fighter sector that won the Battle of Britain. Not far to the north, the thick zone of bomber bases began. Even the Fleet Air Arm acquired an airfield in the near vicinity.

One could say the aim of this study is to test the images and ideas retained from that distant era against the operational prospects for the next two decades, keeping particularly in mind central Europe, the Middle East and Korea — the three most likely locales of large-scale conflict between adversaries well provided with the wherewithal for modern war. There is virtually no discussion between these covers of the air dimension in arms control and phased disarmament. This is not because this subject seems to me entirely otiose or unreal. Nor is it because I reject the now well-established view that deterrence, defence and arms control are the three interacting aspects of the quest for security. Instead, it is largely because the prior issue has to be the changing nature of air power. It is also because the debate about arms control is best pursued in terms of given regional situations projected forward in time-spans of five or ten years. This analysis is at once more global and more long-term in its perspectives.

By much the same token, nothing in the way this text is cast is intended to imply that a major air war is not merely inevitable but already in the offing. In fact, the premises worked from are simply these. There is a risk of war in our generation. Stable deterrence has a central part to play in minimising that risk. Viable capabilities for military operations at the theatre or regional level play a significant part in deterrence.

As regards notation and terminology, the chief thing to say is this. The units of measurement employed in the open literature

Preface

divide remarkably evenly between the metric and the traditional Anglo-American. What is more, there are instances in either case where neatly rounded numbers are used in a very deliberate way. Therefore, these two systems are used interchangeably below. Basic conversion ratios are provided in Appendix B, along with data on financial relativities and certain technical terms. An account of the electromagnetic environment is given in Appendix A.

The phrase 'an order of magnitude' is accorded its formal mathematical meaning — that is to say, a close to ten-fold change or difference. Also, a billion is taken throughout to be a thousand million (10^9). If employed, the term trillion would connote a million million (10^{12}). In other words, French and American usage is followed as opposed to customary British, which would have them as 10^{12} and 10^{18} respectively.

The current NATO practice is to reserve the term 'strike' for nuclear operations and 'attack' for non-nuclear. This has not been adhered to here.

Warmest thanks are extended to the ladies who once again so cheerfully transposed my paleographic longhand into elegant typescript, Mrs Grace Gibbs and Mrs Jill Wells. A very special 'thankyou' is also due to Mr Rodney Melling, former head of Future Projects at the Brough division of British Aerospace, who put many hours of his time into dialogue about the evolving manuscript. Nor would this study have been possible without the assitance so fulsomely proffered by the library staffs of the International Institute for Strategic Studies, the Royal Aeronautical Society and the Royal United Services Institute for Defence Studies. Equally, I am grateful for the advice so freely given by many other specialists in military aviation or other relevant fields. They include officers or other permanent staff of the Luftwaffe, the Royal Air Force, the Royal Aircraft Establishment at Farnborough, the Royal Artillery, the Royal Australian Air Force, the Royal Australian Navy, the Royal Navy, the Royal Swedish Air Force, the Royal Swedish Navy, the Stockholm International Peace Research Institute, Supreme Headquarters Allied Powers Europe (SHAPE), the SHAPE Technical Centre at the Hague, and the United States Air Force. They also include Peter Borcherds, Reginald Bott, Ian Cheeseman, Sir Christopher Foxley-Norris, Richard Garwin, Sir John Hackett, Barry Hughes, Gordon Little, Sir Kenneth Mather, John Moss, Sir Peter Terry, John Waddy and Jim Whitaker. All gave of their expertise generously and always

Preface

without prejudice to whatever wider inferences might be drawn.

Finally, I would like to record my appreciation of two men whose interest and encouragement was invaluable to me as I came to study modern war. The one is the late Sir Basil Liddell Hart. My current disposition, here and elsewhere, to dissent with emphasis from various of his conclusions is no denial of a debt at a deeper level. The other is Laurie Worley (1919–84). He ended World War II as an Acting Squadron Leader and as the commanding officer of 235 Squadron, RAF Coastal Command: a formation operating over the North Sea in rocket-firing Beaufighters. As an old family friend, he told me much — and always with arresting candour — about life and death in wartime skies. To his memory, this book is dedicated.

Neville Brown
January 1986

PART ONE

THE BACKGROUND TO DEBATE

1 THE HISTORICAL SETTING

The Ascent of Air Power

Air warfare is a field in which analysis depends on judgements about a variety of factors which change subtly over time and from place to place. So predictive studies of it must not be shackled to the putative 'lessons of history'. Yet neither can they disregard historical experience. Perhaps the most basic truth to emerge from the past is that it is never possible neatly to distinguish between air power and other modes of military force. During the Battle of Moscow, 1941–2, scores of Wehrmacht tanks were destroyed by Soviet 85 mm anti-aircraft guns. Likewise by 1945, specialist anti-aircraft tanks had come extensively into service. Yet both in Europe and in the Pacific these frequently engaged enemy infantry instead. The same was to happen in Korea and Vietnam.

Quite often, indeed, surface weapons have consciously been conceived as dual-purpose in these terms. As early as 1918, in fact, the Germans were using a *Tank und Flieger* heavy machine-gun against either armour or aviation. Certain naval guns and naval missiles are prominent among the modern instances. Then again, the assignation of weapons as between the air and surface arms may vary from one country to another. Take, for instance, the Nike Hercules, a US medium-calibre missile system solely geared to surface-to-air encounters. Within the US armed forces, it was designated an army weapon. Yet seven other NATO members were to make it an air-force one. Likewise, whether a surface-to-surface missile is seen as 'air power' or 'artillery' may be determined by who is operating it rather than by what it does. Much the same applies to helicopters and so on.

The next lesson to glean is that, in aviation as elsewhere, new genres of weapons are liable to advance to maturity slowly and unsteadily (see Chapter 6). A case in point is the variable-geometry wing: the kind that can be swung forward to obtain high lift at low speeds or swept back to secure low drag at high speeds. France's aerial 'prophet extraordinary', Clément Ader, took out a variable-sweep patent as early as 1890;[1] and by the late 1940s, studies in wind-tunnels and so on were indicating that monoplanes might thus be enabled to achieve an order of magnitude difference between their maximum and minimum speeds. Yet in the next decade, interest waned because of the weight penalties incurred, plus the limits thus imposed on the mode of lateral control, the carriage of external stores, and wing camber. Not until 1964, when the F-111 first flew, was this concept to come into its own (see Chapter 4).

Similarly, the evolution of cruise missiles has not been at all steady. As early as the 1920s, the development began in Germany of the pulse-jet: the simple form of jet propulsion that was to power the V-1 'flying-bomb' (i.e. cruise missile), a weapon surface-launched against Britain in its thousands in 1944. Then came a delay of 15 years before the next big advance, the introduction of several such weapons (principally by the Soviets) for longish-range engagement at sea. After which followed another near hiatus of 15 years or thereabouts.

Much the same applies to the Air-launched Cruise Missile (ALCM), viewed as a sub-genre. In 1943, the Luftwaffe sank several ships, an Italian battleship among them, with a short-range form of ALCM. Then, during the last winter of the war, a complete Geschwader (a formation with some 90 planes) of Heinkel-111s was adapted to deliver V-1s in this way. A further 1,200 V-1s were thereby launched against Britain, usually when their parent aircraft was some 60 km from her coast and at an altitude of 400 m. Granted, merely one in ten reached even the vicinity of worthwhile targets. Still, even that mediocre result could be compared not unfavourably with various early endeavours with free-falling bombs.

Yet not until the late 1950s did even short-range Air-to-Surface Missiles (ASM) appear again; and only in 1961 did any long-range ALCMS enter service, these being the Soviet AS-2 and the US Hound Dog. Then here, too, little further was to happen for 15 years.

Yet sometimes and in certain respects, military air power has developed fast. On most reckonings, it was only in 1903 with the Wright brothers, that powered flight occurred at all, this to an altitude of several yards and across a distance of forty yards. Yet by 1913, an altitude record in excess of 15,000 ft had been registered, with a corresponding range of several hundred miles. By then, too, a top speed in level flight of over 125 mph had been recorded. In the world war that shortly broke, no fewer than 100,000 warplanes were to be built.

Several enduring stratagems made their debut then. Among them was the mass use of tactical air power to check an incipient enemy breakthrough on the ground. Later, its application on the Western Front was analysed by an RAF officer destined to end his career as Chief of the Air Staff. He noted that, as the formidable German infantry offensive of March 1918 gathered momentum, 'there took place one of the most bitter and intensive rearguard actions ever fought by British aircraft, with results the importance of which it is difficult to overstate, and at a terrible cost in casualties'. He averred that, on 25 and 26 March, 'a breakthrough west of Bapaume was averted, and a gap between the French and British around Roye was contained, by a concentration of every available fighter on low-flying action'. True, he doubted whether this remedy could ever work against an armoured breakthrough.[2] But with the right weaponry it might.

From the Western Front emerged, in fact, the doctrinal foundations of air power. Yet, notwithstanding the magisterial work just cited, the apostles of air power did not greatly rely on empirical analysis to shape their view of the future. Instead, doctrine became (and for long was to remain) too abstract to offer much guidance on operational matters. Meanwhile, it separated into two main streams. One was directly complementary to the new notions of mechanised land warfare. The other focussed on the scope that aircraft uniquely afforded for strategic attack on the opposition's heartland. Witness a report made to the British government in 1917 by a committee under Field Marshal Smuts. This warned that 'the day may not be far off when aerial operations with their devastation of enemy lands . . . on a vast scale may become the principal operations of war'.[3] Out of its findings, the Royal Air Force was born.

In the meantime, however, a less Olympian perspective had been adopted by F. W. Lanchester. This British engineer rates not only as an outstanding student of aerodynamics, but as a most stimulating all-round thinker about air warfare. He argued that aerial attacks on the transport network could create 'a virtually impassable zone . . . in the rear of the enemy's defences, a zone varying, perhaps, from 100 to 200 miles in width . . . Thus in the extended employment of aircraft we have the means of compelling a bloodless victory'.[4] What makes so noteworthy this sanguine anticipation of what now is known as 'deep interdiction' is its emanating from someone who insisted none the less that military aviation was no more than a 'fourth arm': the infantry, cavalry and artillery being the first three (see Chapter 13).

In due course, a most pertinent admonition was to come from Major-General J. F. C. Fuller, a leading British apostle of armoured warfare on land. He remarked that were it feasible:

> to devise a portable landing ground the size of a tennis lawn . . . a far greater step towards the abolition of armies and fleets would be accomplished than can possibly be attained by all the tactical experiments now being carried out.[5]

The other side of that coin was that, in the absence of such a remedy, warplanes might be altogether too exposed to immobilisation on their parent airfields.

Still, since consideration of the outlook for interdiction and counter-airfield strikes develops very naturally out of past experience, it seems to me as well to defer a further review of the historical record in these two spheres to Chapters 7 and 9. But we should now look briefly at certain other themes that were to loom large by 1945. Turning then to the maritime sector, it can be said that, even by 1918, the more radical spirits in several navies were becoming alive to the potentialities of aviation. Herein lies the background to the special account taken of the aircraft carrier during the Washington Naval Conference of 1921-2 and a corresponding willingness to get rid of many battleships. Nevertheless, institutionalised inertia prevented this awareness finding full expression in doctrine and programmes until well after 1939. The US and Japanese navies were to progress more than others because (a) their navy (and army) air arms were not overshadowed by independent air forces; and (b) the horizons of the Pacific are exceptionally wide and clear. Yet in the dominant view of 'all the principal navies, sea power still depended on the battleship'.[6]

From 1940 onwards, events were to tell a different tale. Keeping surface fleets at sea while the enemy commanded the skies above proved very costly in the Norway, Greece and Crete, and Malaya campaigns. Also, with the Japanese attack on Pearl Harbor in 1941 and a less weighty one by the Royal Navy against Taranto in Italy in 1940, carrier task forces proved themselves able to cripple an enemy battle-fleet in its home anchorage. Likewise, in the two mightiest and most critical encounters between the Japanese navy and its adversaries in 1942 — namely, the Coral Sea and Midway Island — there were no surface-to-surface exchanges. Virtually all the damage was done by carrier-based aircraft, marginally abetted in the former case by submarines and off Midway by shore-based planes. Obversely, the protagonists of fully independent air power were quick to point out that more than a dozen US carriers plus scores of supporting warships had only been able to direct a few tons of bombs and shells against the Japanese homeland in the final offensive of the war.[7] Island-based heavy bombers delivered several orders of magnitude more.

Another major aspect of the interaction between air and surface has been the aerial projection of ground troops. Assuredly, World War II will always be seen as the classic era of the tactical paradrop. Yet the years in question failed to bear out the bolder of the prior expectations. Take the boast made in 1939 by Major-General Karl Student, the commander of the Wehrmacht's airborne forces: 'In the coming war, I am going to occupy the whole of Sweden with the 7th Air Division and will see to it that Sweden furnishes us with plenty of iron ore.'[8]

In fact, several of the bigger airborne operations were failures. Two in the Netherlands alone were the descent around the Hague by Germany's 22nd Air Division in May 1940; and the Anglo-American attempt, in September 1944, to break across the Maas–Waal–Neder Rhine complex of rivers via the air landings around Arnhem and Nijmegen. Even the successful use of paratroops to spearhead the blitzkriegs of 1940–1 (as in south Norway, inner Holland and Crete) always involved serious losses, not least of invaluable transport aircraft. It could well be argued, indeed, that paratroops were employed economically in offence only when used in small numbers for lightning raids. One instance would be the seizure by just 55 Wehrmacht troops, in May 1940, of the superb Belgian fortress of Eben-Emael. Later German initiatives included the rescue of Mussolini in 1943 and the near capture of Tito in 1944.

One snag about the airborne option was that, while the forces earmarked for it were elite in aptitude and motivation, they were liable to spend much of their time in rearward areas, poised for opportunities that might or might not materialise. Besides which they were unlikely to triumph in assault landings without elaborate preparation and inter-service co-ordination. Think of the errors made during the Allied invasion of Sicily or the host of difficulties that arose out of the Arnhem–Nijmegen descents having been laid on at barely a fortnight's notice. Yet even after thorough staff work, the prospect was always uncertain. As much is evidenced in the confusion so prevalent among the parachute and glider troops landed on D-Day. Moreover, tolerable weather was a critical requirement. So, as a rule, was arrival around dawn.

Then again, the weaknesses in mechanised mobility and firepower, not to mention actual troop strength, of any airborne forces committed usually left them little scope for offensive action after what hopefully had been their surprise arrival. Therefore, this had to take place close to a nodal point in the enemy's routeways. Yet wherever this nodality was apparent to the attackers, it was bound so to be to the defenders. Hence the reception awaiting the British at Arnhem. Likewise the German seizure of the isthmus of Corinth (during the Greek campaign of 1941) did not entrap the retreating Allied armies in quite the fashion intended.

What is more, this contradiction was accentuated if the attack was obliged to secure an airfield as quickly as possible. This would be primarily in order that transport planes could actually touch down to deliver reinforcements that may not have been parachute-trained, together with a full range of materials. Nevertheless, routine air-landings did become one of the war's most innovatory means of support for the extension of offensive action. Such was especially the case in the jungle-dominated landscape of Burma and the South-west Pacific. This was because (a) overground communications were poor and (b) the introduction of battle tanks or other equipment too heavy to move by air was not usually too urgent a requirement.

Still, quite the most awesome exercise of air power during these years was the bomber offensive (the British at night, soon to be joined by the Americans by day) against Nazi Germany. One strand within the endless debate about this campaign has been how far it diminished the Luftwaffe's presence in other theatres, above all the Eastern Front. Until a terrible raid against Hamburg in August 1943, 'as much of Luftwaffe strength was siphoned off to the Eastern Front as could be withdrawn from the West by the application of a brutal yardstick'.[9] Yet it was several months earlier that the Red Air Force had attained what proved to be a general and lasting superiority over the Luftwaffe. It had done so first on the Kuban sector and then during the huge tank battles around Kursk.

However, that most certainly does not mean that no contribution was made to the timing and extent of the swing of aerial fortune in the East by the obligations imposed on the Luftwaffe in respect of the Third Reich itself. After all, the resources being deployed within German borders by 1942 included not only hundreds of fighters but close to 10,000 pieces of anti-aircraft artillery. Prominent among them was the 88 mm, a type of gun otherwise eminently eligible to use in forward areas against tactical warplanes and also tanks. Let us remember, too, how much at risk the Normandy beach-heads might have been, had the Luftwaffe's strength in warplanes in west central France not fallen by D-Day as low as 350. Thirty-five times that number of aircraft were supporting the Allied landings.

But how then could industrial production continue to rise in Nazi Germany into the summer of 1944? Consideration of the morale aspect has been deferred to Chapter 3. So suffice for now to note that rapid industrial growth had been ensured through the previous year by the scope still left for 'taking up slack', as well as by the ruthless exploitation (with comprehensive pillaging and mass deportations) of much of the rest of Europe. If some 25 per cent of all urban dwellings had not been either destroyed or badly damaged in air attack, the profile of output would have climbed still higher: most likely by a good 10 per cent at its peak.

However, it proved for too long too difficult to bomb with sufficient selectivity. For one thing, hundreds of guns might be clustered round high-value individual targets, such as synthetic oil or ball-bearing plants. For another, it was only in the autumn of 1944 that a real breakthrough came in respect of precision attack. It came through the RAF's exploitation of novel electronics in order to strike, first and foremost, at numerous nodes in the German railway network. But even before that, certain particular raids coupled with the cumulative pressure all round had curbed technical innovation severely. Indeed, this effect may have been critical in regard to various 'secret weapons': the menacing new systems for air defence, maritime warfare and so on that were — in certain ominous instances — just becoming operational as Nazism collapsed.

The ultimate recourse to nuclear weapons apart, the raids against the cities of Japan in 1944–5 receive scant attention within all the controversy about strategic bombing. None the less, her extensive and congested urban areas lay wide open to the mass incendiary attacks that became routine once the United States Army Air Force

(USAAF) found how readily erratic winds and zealous air defence could vitiate precision targeting. A single night's bombing in Tokyo in March 1945 killed 80,000 people and left a million more homeless.[10] A deep-rooted urban culture was shattered by such episodes, an irreparable loss that may yet contribute to social instabilities.

The next major air campaign was that waged during the Korean War (1950-3). Undoubtedly, what was by then the United States Air Force (USAF) contributed materially to the progressive retardation and eventual collapse of the armoured thrust made towards Pusan by the North Korean People's Army through the summer of 1950. Yet it achieved far less against the great offensive sustained in the last few weeks of that year by a People's Liberation Army (PLA) which had crossed the Yalu river border from its native China and which was to drive General MacArthur's multinational army (assembled under UN auspices) back into South Korea once more. This is because the PLA incursion was carried out very largely by foot soldiers, advancing well dispersed and well shielded by winter weather.

With the stabilisation of the front across the waist of the peninsula in the summer of 1951, the USAF began Operation Strangle. This interdiction campaign was undertaken primarily to limit the fighting power of the Communist forward troops and hence to influence the armistice talks getting under way in Panmunjom. From the middle of 1952, however, the targeting list was successively extended beyond transportation and logistics to include hydroelectric plants, the whole of Pyongyang (the North Korean capital) and then a quarter of the 20 major dams in the Haeju area, the 'rice bowl' of North Korea. These were attacked between 13 May and 18 June 1953, immediately after the rice seedlings had been transplanted. It has been claimed that this must have been what induced the Chinese and North Koreans to abandon, some time between 7 May and 4 June, a precept they had previously been adamant about: namely, the repatriation of prisoners of war regardless of their individual wishes.[11] Still, this economic pressure is best seen as complementary to a grim warning concurrently conveyed through various covert channels. This was that, unless the negotiating deadlock were soon broken by the requisite Communist concession, the USA would widen the war's horizons and perhaps resort to nuclear release as well.[12] The dambusting signified a degree of ruthlessness requisite for such escalation, a ruthlessness born of two years of military and diplomatic deadlock.

Another facet of the Korean saga that merits some comment is an insistent USAF claim that, in many battles high above the southern approaches of the Yalu, its F-86 Sabres destroyed no fewer than 818 MiG-15s while losing but 58 of their own number.[13] Doubts arise about this and about the related view that overall Chinese losses in fighter aircraft (a good 80 per cent of which would have been MiG-15s) exceeded 2,000, this thanks to a combination of shootings down and training accidents.[14] A good reason to pursue this matter further is that it bears on how enduring a qualitative edge Western air forces may continue to have over their adversaries.

All the same, the most direct evidence is not available for evaluation here. What can certainly be said, however, is that overcounting all too readily occurs in air war, the Battle of Britain affording famous instances. Besides, by no means all the operative factors would have militated against the Chinese. Often they enjoyed a three-fold superiority in numbers in action which should have enabled them to benefit greatly from F. W. Lanchester's square law (see below). Also, their MiGs could be directed from ground stations securely located just beyond the Yalu. Furthermore, the planes themselves were superior in some aerodynamic respects. Finally, if Chinese losses had been all that mortifying, the fighter-aircraft inventory of the fledgling People's Republic could hardly have risen in the course of the Korean campaign from some 400 to 3,000. Nor could the resolve of its pilots have held up as well as it did.

On the other hand, certain considerations do point the other way. Grave deficiencies in advanced training and in-squadron experience obliged the Chinese to rely far too much on close-formation mass attack. Furthermore, the F-86 had the potentially great advantage of a radar-directed gunsight. Never mind that the type then fitted was unreliable and hard to service, which meant that most 'kills' still had to be made without recourse to it.[15] All in all, a good balance may have been struck by a revisionist assessment which puts the MiG loss rate at 1.7 per 1,000 sorties as against 1.3 for the Sabre. It is a differential which implies an exchange ratio overall of just three or four to one in favour of the latter.[16] As it happened, much the same disproportion was to be registered in 1958 in a series of encounters over the Formosa Straits between Chinese Communist planes and Nationalist ones.

In 1956, a military victory in the Israeli and Anglo-French campaign against Egypt was most clearly betokened by Israel's reaching the Suez line. Still, the Anglo-French involvement (not least the virtual destruction on the ground by the RAF of the Egyptian air force) had been essential to the sustaining of enough momentum behind the Israeli overground offensive to ensure success. The 5,000 civilian vehicles pressed into service for it were mostly so ill-suited to desert tracks that, though this drive across the Sinai was only to last five days as events transpired, nearly half of these trucks were thereby immobilised beyond immediate repair.[17]

The next such war began, on 5 June 1967, with co-ordinated pre-emption against 16 Arab front-line airfields by a much improved Israeli air force, a master stroke executed by planes flying at 500 mph below 150 ft. Some 300 Arab planes were thereby destroyed (nearly all while still on the ground) and many vital runways shattered. The rout of the Egyptian, Syrian and Jordanian armies was then just a matter of time. No matter that historic Jerusalem (where the exercise of air power was circumscribed) was hard-contested. Nor that one Egyptian division (commanded by Saad Shazli, his army's chief of staff in 1973) managed to withdraw right across the Sinai, in tolerably good order, from a sector near Eilat.

The perfection of this pre-emptive blow owed much to 'precise timing, hard training, accurate striking power, understanding of their own limitations, and proper intelligence, including an accurate psychological assessment of their opponents'.[18] Important once again, however, were the attributes of Western warplanes and the armament they bore. Let us compare the Mirage III with the MiG-21F, these being the fighters which respectively spearheaded Israel's air arm and Egypt's. The latter machine had a loaded weight barely half that of the Mirage. Correspondingly, its radii of strike action were much shorter; and it also had far less developed electronics for navigation and fire control. Its pair of 30 mm cannon fired at a rate under half that of a similar brace on the French plane. The air-to-air missilry borne by the MiG was effective to two miles, whereas the Matra R.530 on the Mirage could extend to ten.[19] The aerodynamic excellence of the MiG-21 could not compensate for all that.

For similar reasons Israel's air arm was to be hardly less devastating in response to Egypt's 'war of attrition' along the Suez canal from March 1969 onwards. At little cost to itself, it inflicted severe damage and dislocation on the Egyptian army, not least in respect of its air defence. Then in January 1970, the Israelis embarked on deeper penetration, sometimes bombing targets in the very outskirts of Cairo itself in order to try and effect the political collapse of President Nasser.[20] However, his response was a visit to Moscow to obtain more and better surface-to-air weapons — notably, ZSU-23-4 multiple machine-gun on tracks and SAM-3 crew-served missiles. The upshot was a new cease-fire that August. An embarrassment for Israel had been that, given the low heights and high speeds at which her aircraft had to fly over the populous Nile valley, it proved impossible to match the unerring selection of strictly military targets her pilots had almost always achieved before.

In the final analysis, the 'Yom Kippur' war of 1973 was to confirm the West's qualitative advantage. The story of this struggle has been well told many times. Nor has the air dimension been neglected the way it too often has by the historians of other campaigns. All we need to do now is identify a few salient indicators not addressed specifically elsewhere in this study.

Faced with burgeoning evidence of Arab preparations for war, Israel began to consider aerial pre-emption. By 5 October, her air reserves had been mobilised and a major strike planned. Early on the 6th, the Chief of Staff, General David Elazar, urged it be launched at 13.00. Two years later, a USAF study was to surmise that, by such a stroke, up to 90 per cent of the Arab SAM sites could have been destroyed within a few hours for the loss by Israel of less than ten planes.[21] Instead, the Arab offensive came at 14.00 unmolested.

The systematic suppression of air defences by one side or the other was to contribute considerably, albeit patchily, to the subsequent course of events. Thus the air offensive the Egyptians sustained the first couple of days largely took the form of 'shallow interdiction' attacks on forward airstrips and master radars along with camps and local headquarters. Among some deeper strikes was one early on against the airfield and communications centre at Bir Gifgafa, near the Khatmia pass. Meanwhile, on the Golan front Israeli contingency plans to concentrate to begin with on Syria's anti-aircraft network were discarded in order to attack hundreds of Syrian tanks advancing all too neatly echeloned across the bare and narrow plateau. Nevertheless, by 12 October, roughly half the Syrian SAM batteries had been destroyed or badly damaged, while many of the remainder had withdrawn into the environs of Damascus. Likewise, a third of Egypt's 130 SAM batteries were to be wrecked or damaged, 28 by direct air attack and the rest by ground action. Nor is there any doubt that disruption of the relevant SAM belt much facilitated the crossing of the Suez Canal by General Sharon from 15 October.

What is in dispute, however, is how far the dismay and confusion that then began to afflict the 3rd Army beleaguered around Suez reflected the Egyptian prospects overall. It has been argued that they had quickly re-established their SAM network by a switch line. Therefore, any attempt by the Israel Defence Force (IDF) to 'carry open warfare into Egypt would have led it to disaster'.[22] What that view may too heavily discount, however, is the value to Israel of the Emergent Technologies (ET), especially air-to-surface missiles, by then being rushed from the USA. To cite President Sadat's reported message to President Assad on 20 October: 'To put it bluntly, I cannot fight the United States or accept the responsibility before history for the destruction of our armed forces a second time.'[23]

What is undeniable is that the Arab armies had devoted an awful lot of manpower to active air defence. In the Egyptian case, the total is believed to have been 75,000 men, a third of their total army strength. Yet the necessity of this emphasis (in the absence of really effective air cover) was highlighted by the armoured offensive Egypt attempted on 14 October. In particular, one brigade that pushed well beyond the SAM umbrella was virtually demolished from the sky.

Conversely, certain stationary structures survived air attack extraordinarily well. Thus the Syrians directed their air offensive primarily against the Israeli forward defence in the Golan. It was built around 14 *telal*, small hills of volcanic origin. Each was capped by a matrix of barbed wire, concrete, metal sheeting and sandbags, beneath which were bunkers for a garrison of one or two platoons. Several of these strong-points were to last the whole war.

In the meantime, Israeli aerial attempts to foreclose the initial Egyptian surge across the canal had been thwarted. After not a few planes had been lost thereby, the main weight of air attack was switched to the Golan. But on Monday the 8th, much effort was pitched once more against the Suez line, hits being scored on all bar one of the 14 bridges then in place. Yet the eastward flow of men and materials continued uninterrupted; it usually took less than an hour to repair each of the Soviet-made military bridges (see further, Chapter 7). Soon, the Egyptian air force was to fail just as comprehensively against the Sharon bridgehead, despite its having no bridge at all for the first day and a half.

Perhaps the most instructive question raised by this campaign is how far the Egyptians were dissuaded from an immediate advance deep into the Sinai by their apprehension of Israeli air strikes. Ought one simply to say that (a) their national honour had been sufficiently redeemed by an undeniably brilliant crossing of that most awkward Suez waterway; and (b) they did not wish to provoke some wild (and conceivably nuclear?) Israeli or even American reaction, by appearing to threaten the survival of Israel itself? Alternatively, was it rather that General Ahmed Ismail, the Egyptian army commander, discountenanced 'any operation which involved movement beyond his air defences'.[24] A critical view has been that, if the Sinai passes had been seized between 8 and 10 October, the whole of the peninsula would then 'have been liberated with the incalculable political consequences that would have flowed from such a victory'.[25] However, Mohamed Heikal, the author of these thoughts, had been the confidant of Gamul Abdul Nasser. The more pragmatic Cairo of Anwar Sadat was more disposed to seek what it saw as a calculable political consequence, Israel being obliged to withdraw to her 1967 borders within the context of a negotiated peace. In this, it was also at variance with Assad's Damascus.

The special value of the 1973 campaign to the student of air power lies in its being thus close enough to impart some modern electronic perspectives, yet distant enough to obviate the temptation to draw inferences too specific about, say, SAM-mobility or SAM-suppression. The Vietnam war overlapped it but lacked some forms of advanced electronics almost throughout. Indeed, this and the terrain help explain why 14 million tons of ammunition had to be expended by the USA and her allies.[26] Just over a half of this was air-delivered. So the total for air ordnance alone was seven times the corresponding tonnage for the Korean war and three times that used in the Anglo-American bombing offensive against Nazi Germany. Added to which a leading role was played by aircraft in environmental modification by more esoteric means. Among them was the defoliation of forests. So was the seeding of rain clouds, presumptively to make them more opaque to anti-aircraft radar as well as to try and deluge jungle trails below.[27]

The Falklands war offers us certain particular lessons, at least on a tentative basis. Considered overall, however, it is a typical example of an atypical conflict. The same goes for Israel's invasion of the Lebanon in 1982. Here the extreme width of the quality gap between the main protagonists was savagely highlighted by Israeli aircraft shooting down 80 Syrian warplanes for the loss of but one or two of their own number. Granted, this gap would have been less evident if the Arab surface defences had been stronger. As it was, both the PLO and the Syrian garrison in the Bekaa valley had far less adequate surface-to-air cover than such Soviet-supplied forces might normally boast nowadays. For instance, the Syrians had never properly hardened their SAM sites in that pivotal part of Lebanon, this apparently for fear of provoking Israeli pre-emption.[28]

Irrespective of the singular features of particular episodes, however, the modern Middle East is an amazingly rich and diversified source for the assessment of air power and the consequences of its application. Let us take a few more examples. Egyptian aircraft dropped poison gas in the Yemen in the mid-Sixties. So did Iraqi planes and helicopters in the Gulf War in 1984.[29] At the start of the 1967 campaign, Israeli aircraft savaged the *Liberty* — a US naval surveillance vessel. Shortly afterwards, King Hussein lent his royal authority to talk of Anglo-American air intervention on Israel's behalf, an allegation Radio Biafra was soon to emulate in similar desperation in Nigeria's civil war! One might add that the great era of aerial hijacking began in the aftermath of this June War, with the Middle East becoming one of the most affected zones. The two most spectacular outcomes were the civil war in Jordan in 1970 and the rescue at Entebbe in 1976. Fearful for their nuclear complex at Dimona in the Negev, the Israelis shot down a Libyan airliner astray over the Sinai early in 1973, over 100 civilian passengers thus being killed. They shattered Iraq's key nuclear facilities in an air strike in June 1981. Their clandestine airlifts have included running supplies to Iran, especially early on in the Gulf war.[30] As of late 1985, one doubts whether the emigration of Falasha Jews from Ethiopia to Israel can be completed without some military air involvement. Then again, the great majority of Israel's punitive strikes in the Lebanon have been air-delivered, and their insistence on continued air surveillance is among the specific obstacles to a Lebanon disengagement pact. Without air power, the Soviet intervention in Afghanistan would have been doomed from the start.

In fact, one general inference to be drawn, even from such a skeletal overview as this, has been that in twentieth-century war defeat will almost always be avoided (and outright victory very likely gained) by the side that has secured air superiority. Indeed, a more comprehensive perusal would most probably show that virtually the only exceptions concern insurgency warfare.

Still, that may not be the most instructive lesson to glean from the historical record. Rather this may be that the exercise of air power has been governed by geography more than is usually appreciated. By geography, one does not mean just the terrain. The width of front is also most relevant, especially as judged in relation to the levels of force deployed. Likewise, the depth of the theatre in question affects the scope afforded air power. So do the human and economic resources disposed across the surface.

A rider may be added to this. The interaction between geography and air power may influence not only the course of armed conflicts. It may also be a critical determinant of whether they can start at all. In his famous numerical study of the incidence of warfare, the mathematical meteorologist and Quaker pacifist, Lewis F. Richardson, addressed this point. His view was that the actual occurrence of wars has been much less than might have been feared, given the opportunities presented by geographical adjacency. But he went on to caution that those same opportunities had been expanded greatly by the modern development of naval and air power.[31] In a region as close-knit as Europe, it would be hard to demonstrate that this effect has been all that pronounced, the blitzkrieg era notwithstanding. But the Middle East is surely a clear example of a zone in which resort to armed violence has been rendered more feasible, and sometimes more imperative, by the existence of air power.

The Evolution of Analysis

As always, the future of air power is largely being determined by the viability or otherwise of tactical warplanes. As was remarked in 1967 by the then Controller of Aircraft at Britain's Ministry of Technology, this problem had first exercised air staffs 'in the late 1940s, once they had realised the potential impact of the guided missile in all its applications'. The maritime environment apart, the outlook was already 'debatable' for tactical aviation though 'with battlefield support probably surviving longest'.

However, the gestation of a new genre of weapons tends to be longer than radical thinkers anticipate, not least because of engineering bottlenecks and fiscal curbs. Sometimes, too, insufficient allowance is made for improved technology and tactics in the domain of warplanes as such. More specifically, the problem of low-level interception has proved more intractable than expected. Acting together, these several sources of error 'could well have led one to predict a demise of the manned combat aircraft in the tactical role at the latest by the mid-1960s, had one been required to look so far ahead'.[32]

In fact, all that really happened in the interval was that professional endorsement of the warplane became lower key and more contingent. An essay by a British author published in 1966 allowed it a significant role in limited war in the 'immediate', though maybe not the more distant, future. The critical variables identified were improvements in anti-aircraft defence and in airfield attack, along with the direct substitution of aircraft by missiles. Yet once again, maritime operations were made an exception. Also, the warplane was explicitly conceded an extra lease of life in countries unable 'to develop guided or ballistic missile systems of their own or develop nuclear weapons'.[33] A Soviet assessment published in the same volume stressed the endless oscillation to be observed historically, in spheres like air warfare, between the respective fortunes of Attack and Defence.[34]

Then, in 1968 came the third and final edition of a synoptic review of modern war undertaken by a syndicate of young staff officers in the Soviet army, under the chairmanship of the late Marshal Sokolovsky. Informed by a customary Soviet emphasis on the human factor, it summed up the tactical air picture in this ambiguous fashion:

> A considerable portion of the missions formerly executed by frontal bombers are also beginning to be handed over to operational tactical missiles. The arming of bombers and fighter-bombers with various classes of missiles enables them to operate successfully on the battlefield . . . in support of ground forces, especially in zones with a weak anti-aircraft defence. Furthermore there are many specific missions, for example the destruction of moving targets, which can be executed more successfully by bombers or fighter-bombers than by missiles.[35]

Meanwhile, the full text is studded with allusions to the supposedly regular alternation, throughout military science, between given weapons and their antidotes.

That same year, a French military opinion was that the offensive use of aircraft, and of transport aircraft in particular, may become problematic. Thus the preliminary battle for 'mastery of the air', the great theme of 1945, would be replaced by one for the 'freedom of the air'. For one thing, Electronic Counter-Measures (ECM) 'could permit a reasonable survival of aircraft in flight at least in one sector and for a given time'. In electronics as a whole, the 'truly enormous experiment which will take place at the beginning of the conflict will be decisive in this field and could have very surprising results'.[36]

In this same volume, a British specialist on aeronautical mathematics wrote of the progress then being made in low-level air defence. True, this was as yet being circumscribed by the 'terrain screening the target. But the advent of a cheap lightweight anti-aircraft guided missile that could be used in large numbers by the soldier in the field would weigh heavily against the strike aircraft operating over enemy territory'. Even without such opposition, the difficulty of identifying ground targets would make 'increased development and use of the area attack weapons . . . the only foreseeable trend'. A putative alternative, the 'stand-off' missile, was thought unlikely to be viable, save with nuclear warheads.[37] Again, an exception could have been made of the maritime environment.

Finally, in 1972, a study conducted at the International Institute for Strategic Studies (IISS) concluded that 'while the increasing effectiveness of SAM systems has, for all practical purposes, disposed of the threat of a medium altitude attack' the probability was that the task of countering the low-flying aircraft, whether with ground weapons or interceptor aircraft, will remain a formidable one for well beyond the present.[38] This was partly because of the technical obstacles to the development of an airborne radar with an adequate 'look-down' capability.

However impressionistic these various judgements may appear, each and every one will have been influenced by the extension into this sphere of systematic scientific analysis, both deductive and inductive. Veritably a classic exercise in deduction (the application to particular situations of inferences drawn from general premises) was the essay published in 1916 by F. W. Lanchester. Quite the most celebrated of its formulations was the square law: 'the fighting strength of a force may be broadly defined as proportional to the square of its numerical strength, multiplied by the fighting value of its individual units'.[39] In other words, a marginal increase in number is liable to be more consequential than a marginal gain in quality.

In the ultimate, this logical validation of the 'principle of mass' must hold good for all patterns of military force. However, its impact on events will be most readily manifest when the opposing sides have free access to each other. Where, as in most historical land campaigns, geography imposes constraints, the relationship between numbers and effective strength is more nearly linear.[40] Lanchester himself felt best able to demonstrate his deduction through an interpretation of the Royal Navy's battle plan at Trafalgar in 1805.[41] Still, his great concern was to address contests between rival air fleets.

Alas, it has to be said that comparison of quality is nowadays far harder than in 1916, not that it was so very easy then. How does one ascertain differentials in electronic performance? And how can one express them arithmetically?

Some carry a lot further their critique of the Lanchester legacy. They perceive a tendency to gauge quality merely in terms of weaponry presented. They regret indifference to how the attritive capacity of a given force at a given time is influenced by the strength of the opposition. They apprehend, too, a disposition to see damage infliction just as a matter of physical attrition, demoralisation and disorganisation thus being discounted.[42] All the same, the principle of mass is bound to remain considerably relevant to military aviation.

As World War II progressed, the main thrust of methodological development shifted well towards the form of inductive enquiry that came to be spoken of as Operational Research or Operational Analysis, by which is meant the collection and analysis of numerical data in order to identify patterns, correlations and trends in actual war, with policy prescription in mind. That such processing then took place in the Soviet Union, too, is indicated by the arithmetic relationships and tendencies appertaining to those years so regularly recorded in the pages of Soviet military literature.

All the same, little or nothing has yet been published about the structural development of this discipline within the USSR during those wartime years. Conversely, a lot has been written about Anglo-US recourse to it, not least the lead given by the front-line commands of the Royal Air Force: Fighter, Bomber and Coastal.

Most of the problems that readily invited this approach concerned air warfare as such or else the interface between that and maritime conflict.

Masterminding the interdiction campaign against the French railway system in the spring and early summer of 1944 was certainly the paramount case in point. Among other matters addressed were the best patterns in which to cluster anti-aircraft guns around radar sites in defence of London; evaluating the performance of coastal anti-aircraft guns; how much benefit accrued from Combat Air Patrols (CAP) above oceanic convoys; whether to have the depth charges being air-delivered against U-Boats set to explode closer to the surface; making visual surveillance from Long-Range Maritime Patrol (LRMP) planes as synoptic as possible; and exploiting the principle of mass, as far as was allowed by contingent circumstances, in the employment of strategic bombers.

Judged as the fruits of scholarly enquiry, some of the conclusions reached appear, in retrospect, painfully banal. Yet the reality was that only such systematic presentation of evidence and inference, underscored by the high prestige of certain of the scientists involved, was able to carry the day against bureaucratic inertia, obfuscation, obscurantism or outright resistance. Thus, records previously thought to reveal that anti-aircraft artillery was twice as effective in coastal locations as when it was employed deeper inland were persuasively reinterpreted to show that they did not derive from such ambient factors as better radar reception low above sea surfaces. Rather they were due to the impossibility in that environment of checking gunners' claims against the number of crashed enemy aircraft subsequently discovered. A more than twofold exaggeration in the initial reporting of enemy losses was, in fact, the norm throughout British air defence early in the war.[43] Nor has such a disposition been absent from the more recent past, as witness the Korean War experience above.

Investigations broadly akin to these took a big leap forward in the USA with the appointment of Robert McNamara as Secretary of Defense in 1961. What he promoted, above all else, was the routine inclusion of financial costings and evaluations. In the great majority of cases, the studies conducted during World War II had been informed only by vague reckonings of comparative cost-benefit. For one thing, they had preceded the full flourishing, in other spheres, of the science of econometrics. For another, the systems they were concerned with had already been fully developed and largely procured.

Not that this revamping of operational analysis was ever to make a truly substantive contribution to the management of any aspect of the arms race. All else apart, a great deal of its model-building was to remain amazingly crude. Take, for example, an expanded version of a 'highly aggregated ground-air conflict computer model' first devised for the USAF by its protégé, the Rand Corporation in the early 1950s. Even after the said revision, 'no distinction is made between armored or mechanised divisions and infantry divisions' (Emerson, 1973, p. 2).[44] The supply system 'is primarily a rail network' (p. 2). 'No consideration is given to the effects of damaged airfield facilities' (p. 2)! The Forward Edge of the Battle Area (FEBA) moves as an entity, the rate duly being 'the average movement of the entire theatre front' (p. 4). There are but three types of aircraft on each side. 'Airfields are not subject to being overrun in the ground battle' (p. 6). As a general rule, 'Attacks against aircraft in shelters are presumed to destroy aircraft in any occupied ones that are hit, but to leave the shelters intact for subsequent re-use' (p. 9)! Each army division is assumed to have just 'two types of mobile nuclear fire-units' (p. 13). Finally, even though the nuclear threshold is destined to be crossed, the total number of air-delivered nuclear weapons employed per day is presumed to be so small that 'no specific requirement is imposed for aircraft or sorties for delivery' (p. 15). With postulates as crude as this, the use of a computer seems an otiose refinement. Nor is this by any means an exceptional example of this kind of arbitrariness.

Moreover, doubts on this score are accentuated by the extensive philosophical overlap between the McNamara approach to systems analysis and the decision-making behind the Indo-China involvement. My own experience of the Vietnam war was confined to one short visit, as a journalist, in 1966. But even that was enough to convince me that the overriding disposition of the US analysts (though not the field troops or aircrews) was to attach an almost absolute importance to whatever was neatly quantifiable. The tendency was to highlight military achievement as opposed to political; and also to gauge that achievement more in terms of cumulative outlays (tonnages of bombs, leaflets and so on) than of any tactical gains.

Besides, the inferences about development and procurement that emerged out of systems analyses informed by the precept of comparative cost-benefit proved — in several salient instances — unable to prevail in the political arena. Thus, in 1963, Secretary McNamara testified to the Senate Armed Services Committee as follows:

> Now, I think everyone would agree in the Defense Department . . . that strategic launch platforms for free-fall bombs will be obsolete in the 1970s, without any question . . . So the . . . alternatives are missiles launched from the sea, missiles launched from the land, or missiles launched from the air, and of those three modes of launch the missiles launched from the air are by far the most complex missiles, likely to be the most expensive and unreliable, and certainly would require the most advanced development.[45]

Yet in 1985, we find that the official policy in Washington (not to mention Moscow) is to keep the strategic bomber in service for decades to come.

The plain truth is that the operational analysis of aerial conflict post-1945 has too often lacked enough incision and pertinence to weigh against pre-existing currents of opinion and interest groups. No matter that, at any rate prior to the recent burgeoning of electronics, air warfare has seemed to lend itself to logical and numerical evaluation better than has combat on land. This has been partly because its main instruments have tended to be discrete and measurable; and their application less circumscribed by geography. Yet it has also been because the human reactions of the aircrews themselves have appeared relatively predictable, in that they are those of highly trained and indoctrinated men responding within a detached and esoteric machine environment. But what one should mean by 'relatively' is the crux of the matter, not least as regards the uncertain thresholds of human stress.

Sortie and Loss Rates

One index of air power that is never examined as comprehensively as it should be is the number of sorties men and planes can be expected to fly in the course of a full fighting day. A dramatic illustration of the impact differences therein can make, through the Lanchester square law and all that stems from it, is afforded by the 1967 Arab–Israel war: 'the exceedingly fast turn-round of the Israeli aircraft — some sources have quoted seven to ten minutes — meant that the Israeli Air Force could fly up to eight one-hour sorties in a day'.[46] Yet it seems the Egyptians had reckoned on its only managing four. Hence President Nasser's astonishment at the streams of enemy aircraft presenting themselves; and King Hussein's assertion of Anglo-US aerial involvement.

Still, one thing to guard against is talking about the sortie frequency registered by aircrew and that registered by the machines they man as if the two always coincide. Granted, it was normal in most air forces in World War II to align one crew with one plane on a quasi-permanent basis, a prime motive being to ensure that every pilot was familiar with the idiosyncracies of his machine. But this practice has been widely discarded, the number of aircrews now regularly well exceeding the number of aeroplanes, except in aircraft carriers (see Chapter 3). The Danish F-16 programme, which seems quite typical of modern Western practice, allows a ratio of 1.35 to 1.0.[47]

As a rule, attention is primarily paid to sortie rates per plane in operational service. However, this estimation may be complicated by a proportion of a given air fleet being refitted or in reserve. Although in August 1940 RAF Fighter Command was being stretched almost to breaking point, it still had a third of its total inventory of Spitfires and Hurricanes in maintenance lines.[48] As it happens, exactly the same applies today to the B-52 echelon of Strategic Air Command.

Another distinction to preserve is that between air defence and intrusion. RAF experience during the Battle of Britain was that, whenever squadrons sought to launch four or more interceptions or Combat Air Patrols (CAP) between dawn and dusk, the strain they thereby imposed on themselves soon became unbearable. So two a day was made the norm.

All the same, strikes deep inside hostile air space usually tax those concerned more severely. Also, such a mission takes relatively long to prepare, never mind that the burden of servicing is now tending to ease all round. A hundred missions was the total averaged by F-111s operating out of Thailand against the Red River basin of North Vietnam during six crucial winter months in 1972–3.[49] That might well be typical of a deep interdiction campaign, sustained over time.

When instead the purpose is Close Air Support (CAS), the

distance of the air bases or airstrips from the land battle in progress assumes a special if not decisive importance. Thus the RAF claims that, dispersed well-forward, its Vertical Take-Off and Landing (VTOL) monoplanes, the Harriers, can sustain six sorties per aircraft per day. Likewise on the Eastern front in World War II, forward-basing by the Luftwaffe and Red Air Force respectively did facilitate the generation at critical junctures of exceptionally high rates. Even Soviet heavy bomber squadrons generated five sorties a night against German troops retiring from Leningrad in 1944.[50] Sometimes, particular Stuka dive-bombers had been over Sevastopol 12 to 18 times within 24 hours in 1941–2, albeit as part of a thoroughly indiscriminate bombardment of this beleaguered city.[51] During the blitzkrieg against Western Europe in 1940, six sorties a day had been a Luftwaffe norm.

One determinant, especially in relation to the allocation of resources for CAS, is how long a conflict may last and how its intensity may vary over that time. A while back, the indications were that NATO air staffs had finally come to regard two or three attack sorties a day as altogether too few for the dynamism — and hence the brevity — of a future 'round-the-clock' war in Europe. As was pointedly observed in 1977 by the Assistant Director (Air Warfare) in the US Directorate of Defense Research and Engineering:

> The traditional sortie rates must be tripled and sustained for significant periods . . . Such sortie rates are hardly possible with large formations which combine attack, fighter escort and defence suppression orchestrated by an airborne controller. Instead, many small flights on numerous tracks appear to be a more manageable and effective operational concept.[52]

No doubt, though, this inference was also driven by USAF recognition that, in any case, the former mode was less well adapted than before to the penetration of hostile air space. These days, the further point would be made that airfields under attack cannot generate high enough sortie rates on the basis of large formations.[53]

Meanwhile, more attention has been paid to enabling squadrons to surge to abnormally high sortie rates for the first several days of a conflict. Even so, a 1981 report did not find the 'surge' rates at all pleasing up to that time, except as compared with certain absurdly low previous norms. The F-16 Fighting Falcon could rise to 5.1 sorties per plane per day as against a wonted 0.75. But the F-15

Eagle, being primarily an air superiority fighter, went from 0.84 only to 3.3. The A-10 Thunderbolt 2 (Close Air Support anti-armour machine) went from 1.0 to 6.0, though the more deeply striking A-7 Corsair only from 0.85 to 3.3. Interdictions by the F-111D rose from 0.43 to 2.0.[54] Doctrine and mission specifics apart, the factors that determine a sortie profile over time include manpower and infrastructural support; the availability of spares; and, not least, the serviceability of the machines themselves. Illustrating the influence of the last factor is Northrop's claim that the greater reliability and easier maintenance of their privately developed F-20 Tigershark could permit 40 per cent more sorties per day than the F-16, say, could manage (see Chapters 12 and 13).[55]

Given all the variables, comparisons with adversary air arms are singularly difficult. Arguably, for instance, enough can be deduced about the Warsaw Pact in central Europe to warrant the expectation that its sortie rates per plane per day would not match NATO's. The criteria here alluded to are less flying training at night or in bad weather; less use of the ground-based simulator as a training aid; less air-force manpower per plane; aircraft serviceability; and more recourse to rough-field operation, which tends to involve dispersion with an added stretch on resources. Nevertheless, a 1976 analysis by a staff member of the Arms Control and Disarmament Agency in Washington was crediting the Warsaw Pact with a rate 20 per cent higher.[56]

Sortie rates off carriers are lower than from land bases, almost regardless of the specific circumstances. Distances to target or patrol circuit sometimes contribute to this. But a major constraint is the handling of aircraft on board. All commentators agree that the Royal Navy did well to average two CAP or interception sorties a day with their Sea Harriers and Harriers during the most crucial five days of the Falklands war.

Another vital parameter of air power that is always hard to assess comprehensively is the rate of loss in operational sorties. Suffice for now to make some general comments. To start with the most apparent, the lower limit of losses is effectively zero. As to what the upper may lately have been, a prime indicator is the kill rate of two in three or thereabouts, per target engagement and in ideal conditions, claimed for a succession of surface-to-air systems from the 1960s onwards. Allowing for a return pass, that could suggest a rate of attrition close to 90 per cent. In practice, however, this is rarely achieved. All else apart, a war situation is never 'one-to-one': an individual anti-aircraft piece versus an individual plane.

Towards the end of World War II, the average attrition rates endured by aircraft at particular times and places almost all ranged between a minimum of 1 per cent and a maximum of around 35 per cent. Within these limits, placings were determined by force densities, electronic technology gaps, the Lanchester square law, the element of surprise, sheer chance and so on.[57] Moreover, much the same has applied in recent times. One per cent attrition was the average in the Indo-Pakistani war of 1971.[58] The Israelis report that the loss rate they suffered throughout was 1.4 in the 1967 war and 1.0 in the 1973, the overall level having been forced up on the first occasion by the application of its Magister light strike-trainers to Close Air Support (CAS) (see below, p. 243). Obversely, the Egyptian rate of loss in the many hundreds of close support sorties launched in desperation in the closing stages of the 1973 conflict is believed then to have risen to over 20 per cent.

At the climax of the Falklands campaign (21–5 May) Britain's Harriers and Sea Harriers flew some 300 sorties as compared with 200 by Argentinian fighter-bombers operating from the mainland. One Harrier was shot down, while a Sea Harrier crashed on take-off. At least 19 (some analysts say over 30) of these Argentinian machines were shot down. So we can reckon on an attrition rate of 15 to 25 per cent for those 'hostiles' which actually entered the British air defence zone.[59]

Within such a spread, the distribution of results on specific occasions will tend to have a strong 'left-handed' skew. This is on a standard xy graph: the x axis measuring attrition rate and the y the rate of occurrence at each value of x. In other words, the peak rate will be well towards the lower end of the total spread — maybe 6 or 8 per cent on current showing. Such a bias in fact occurs in many statistical patterns of a broadly analogous kind. Nor is this surprising. Marginal shifts in the balance of factors are liable to have a more dramatic effect on the higher side of the spread. So the xy curve will be much extended there.

But to all the data on warplanes shot down ought to be added a penumbra of ones damaged in flight. Thus in the Battle of the Ruhr (5 March to 12 July 1943), RAF Bomber Command suffered a total loss rate of 4.7 per 100 sorties but, in addition, a damage rate of 16 per 100.[60] However, most commentators would say that the ratio of planes damaged to those destroyed will be less these days because of (a) the sheer deadliness of modern ordnance and (b) the machines in question often being hit when flying too fast too close to the ground to be able to accept any loss of control. However, helicopters are quite another matter. In their case, the ratio of damaged to destroyed can be very high.

Another caveat to enter is that the spectrum of probability delineated above does not extend to a nuclear situation. Ignoring for the moment a much accentuated risk of pre-emptive destruction on base, one can allow that attrition rates might be low in a tactical nuclear exchange because (a) the respective defence environments would have been degraded considerably by the time this threshold was crossed, (b) nuclear bursts would accelerate this degradation, and (c) it might be difficult to use nuclear interceptor missiles above friendly territory.

Still, the fact of the matter is that such calculation is rarely done apropos theatre nuclear war. On the other hand, the presumed penetration prospects of manned bombers have prominently figured in the debate about strategic nuclear deterrence. Early in the 1970s, one USAF source was suggesting that effective penetration by its bombers of the USSR could be as high as 85 per cent.[61] But that bold claim may have rested heavily on the premises that (a) strategic missiles would pave the way, (b) the USA retained a significant margin of superiority in the EW field, and (c) Soviet defences lacked an adequate capability for 'look-down, shoot-down', the interception from above of intruders flying low. To all of which has to be added primordial optimism, a human attribute that aviators usually possess in good measure.

There are two other issues which will be pursued later in the book (see Chapters 7 and 13). The first is the relative powers of anti-aircraft guns and SAMs; second, the vulnerability in flight of cruise and ballistic missiles. On the latter, the evidence is desperately patchy. What is more, the predictive interpretation of it is too sensitive to subsidiary judgements, both intrinsic and extrinsic. Thus *a priori* speculation about the survival to target of the cruise missile is ruled out because the exact connotations of its tight contour-hugging are far from certain. Similarly, it is hard to say how far the very rapid approach of a ballistic missile is offset by its following a most visible and predictable parabolic flight path.

Experience with one cruise missile, the above-mentioned V-1, does exemplify very well a most persistent feature of air warfare as a whole. This is that, faced with the onset of actual fighting, surface-to-air defences do not approach their notional capabilities for several hours or even days, assuming, of course, that they ever do. Though the same applies to intruding flying machines, it does so to a lesser extent. To put the argument no higher just for now, the former has to react to the latter more than vice versa.

Thus out of the 16,000 V1s launched against England and the Low Countries in 1944-5, nearly 7,000 were to be destroyed by fighters, anti-aircraft guns or barrage balloons. Yet the first waves suffered only 2 per cent attrition. Within a week, however, near to 50 per cent was being registered. Towards the end of the attacks on England, anti-aircraft batteries were shooting down up to 80 per cent of those V1s that crossed their sights.[62]

Although British warplanes launched 1,500 low-level sorties against Egypt during the Suez crisis of 1956, only eight were shot down. For what it may be worth, however, the ability of the Egyptian artillerists to hurt, distract or deter did peak, however subtly, on the third or fourth day. But more indicative is the Linebacker 2 offensive conducted by the USAF against Hanoi and Haiphong in December 1972 and January 1973. Some 700 high-altitude sorties were made by B-52 heavy bombers, 15 of these planes being lost and 6 more damaged. Here again, the most severe attrition (some six B-52s shot down) came on the third to fourth day. Before that, the North Vietnamese capabilities were not galvanised sufficiently. Afterwards, the suppression of them took over.

A double exception to this pattern is afforded by Israeli experience in the 1967 and 1973 wars. For on both occasions, their attrition rate was at a peak level of 4.0 per cent during the first two days. Still, this pair of anomalies can be explained. In the first day or so of the 1973 campaign, Israeli air squadrons deliberately ran exceptional risks (a) to check Syrian armour on the Golan and (b) to try and break the Egyptian bridges over the Suez Canal. Then as regards 1967, their Magister light strike-trainer force had suffered attrition rates several times the then Israeli average when freely used very early on against Arab field formations that had yet to disintegrate. One has to add that Israeli air losses were bound to fall away after the first day or two in 1967 because by then the Arab air defence was not just underreacting, it was in ruins. In fact, of course, that June war affords the supreme illustration of the almost pathetic exposure to attack of airfields, above all in the first hours of conflict.[63]

References

1. Oliver Stewart, *Aviation: The Creative Ideas* (Faber & Faber, London, 1966), p. 30.
2. J. C. Slessor, *Air Power and Armies* (Oxford University Press, London, 1936), pp. 105–6.
3. *Air Power Historian*, 3 July 1956, pp. 148–53.
4. F. W. Lanchester, *Aircraft in Warfare, the Dawn of the Fourth Arm* (Constable, London, 1916), pp. 187–8.
5. J. F. C. Fuller, *On Future Warfare* (Sifton Praed, London, 1928), p. 202.
6. Arthus Hezlet, *Aircraft and Sea Power* (Peter Davies, London, 1970), p. 135.
7. See 'The Myth of the Aircraft Carrier' in Alexander P. de Seversky, *Air Power, Key to Survival* (Herbert Jenkins, London, 1952), ch. VI.
8. Quoted in Maurice Tugwell, *Airborne to Battle* (William Kimber, London, 1971), p. 40.
9. Adolf Galland in Eugene Emme (ed.), *The Impact of Air Power* (Van Nostrand, Princeton, 1959), p. 293.
10. For eye-witness accounts at ground level see *Cries for Peace* (Japan Times Ltd, Tokyo, 1978), ch. 3.
11. 'The Attack on the Irrigation Dams in North Korea', *Air University Quarterly Review*, vol. VI, no. 4, Winter (1953–4), pp. 40–61.
12. David Rees, *Korea, the Limited War* (Macmillan, London, 1964), ch. 20.
13. M. S. Knock, *Encyclopedia of US Air Force Aircraft and Missile Systems* (The Office of Air Force History, Washington DC, 1978), vol. 1, p. 63.
14. R. M. Bueschel, *Communist Chinese Air Power* (Praeger, New York, 1968), p. 26.
15. W. W. Momyer, *Air Power in Three Wars* (Lavalle, Washington DC, 1978), p. 115.
16. Maurice Allward, *F-86 Sabre* (Ian Allan, London, 1978), p. 46.
17. See the author's *Strategic Mobility* (Chatto and Windus, London, 1963), p. 64.
18. Robin Higham, *Air Power, A Concise History* (Macdonald, London, 1972), p. 229.
19. See the author's *European Security, 1972–80* (Royal United Services Institute for Defence Studies, London, 1972), p. 74.
20. A. Schlaim and R. Tranter, 'Decision Process, Choice and Consequence', *World Politics*, vol. 30, no. 4, July (1978), p. 483.
21. S. J. Rosen and M. Indyk, 'The Temptation to Pre-Empt in a Fifth Arab–Israeli War', *Orbis*, vol. 20, no. 2, Summer (1976), pp. 265–85.
22. Anthony Farrar-Hockley, *The Arab-Israel War, October 1973, Background and Events*, Adelphi Paper 111 (International Institute for Strategic Studies, London, 1974), p. 30.
23. Quoted in Mohamed Heikal, *The Road to Ramadan* (Collins, London, 1975), pp. 238–9.
24. Farrar-Hockley, *The Arab–Israel War*, p. 26.
25. Heikal, *The Road to Ramadan*, p. 215.
26. Raphael Littauer and Norman Uphoff (eds), *The Air War in Indo-China* (Beacon Press, Boston, 1972), Statistical Summary.
27. *Ecological Consequences of the Second Indo-China War* (Almquist and Wiksell for SIPRI, Stockholm, 1976), ch. 3, sec. 4.
28. See the author's Inaugural Lecture, *Silver Wings in the Twilight* (University of Birmingham, Birmingham, 1983), Appendix A.
29. *New Scientist*, 22 March 1984.
30. *Strategic Survey, 1981–82* (International Institute for Strategic Studies,

London, 1982), p. 90.
31. L. F. Richardson, *The Statistics of Deadly Quarrels* (Stevens and Co., London, 1960), p. 294.
32. Christopher Hartley, 'The Future of Manned Aircraft' in *The Implications of Military Technology in the 1970s*, Adelphi Paper 46 (International Institute for Strategic Studies, London, 1968), pp. 28-37.
33. Martin Edmonds in John Erickson (ed.), *The Military-Technical Revolution* (Pall Mall, London, 1966), pp. 170-86.
34. N. Talensky, ibid., pp. 219-28.
35. H. F. Scott (ed.), *Soviet Military Strategy* (Macdonald and Jane's, London, 1975), p. 253.
36. Andre Beaufre, 'Battlefields of the 1980s' in Nigel Calder (ed.), *Unless Peace Comes* (Allen Lane, London, 1968), pp. 15-25.
37. Andrew Stratton, 'Contests in the Sky' in ibid., pp. 69-97.
38. Trevor Cliffe, *Military Technology and the European Balance*, Adelphi Paper 89 (International Institute for Strategic Studies, London, 1972), p. 21.
39. *Aircraft in Warfare, the Dawn of the Fourth Arm*, p. 26.
40. D. L. I. Kirkpatrick, 'Do Lanchester's Equations Adequately Model Real Battles?', *The Journal of the Royal United Services Institute for Defence Studies*, vol. 130, no. 2, June (1985), pp. 25-7.
41. *Aircraft in Warfare, the Dawn of the Fourth Arm*, p. 66.
42. J. N. Merritt and P. M. Spreu, 'Negative Marginal Returns in Weapons Acquisition in Richard G. Head and Ervin J. Rokke (eds) *American Defense Policy* (John Hopkins University Press, Baltimore, 1973), pp. 486-95.
43. P. M. S. Blackett, *Studies of War* (Oliver and Boyd, Edinburgh, 1962), pp. 211-12.
44. Donald E. Emerson, R-1242-PR, *TAGS-V: A Tactical Air-Ground Warfare Model* (Rand Corporation, Santa Monica, 1973).
45. William W. Kaufmann, *The McNamara Strategy* (Harper and Row, New York, 1964), p. 228.
46. Geoffrey Kemp, *Arms and Security: The Egypt-Israel Case*, Adelphi Paper 52 (International Institute for Strategic Studies, London, 1968), p. 6.
47. *Aviation Week and Space Technology*, vol. 114, no. 24, 15 June 1981, p. 84.
48. Chester Wilmot, *The Struggle for Europe* (Collins, London, 1965), ch. 1.
49. Bill Gunston, *F-111* (Ian Allan, London, 1978), p. 61.
50. B. A. Vasil'yev, *Long-Range, Missile-Equipped* (Military Publishing House, Moscow, 1972), ch. 2.
51. Alfred Price, *Pictorial History of the Luftwaffe, 1933-1945* (Ian Allan, London, 1969), p. 33.
52. Charles E. Meyer as quoted in *Aviation Week and Space Technology*, vol. 106, no. 6, 7 June 1977, p. 24.
53. 'USAF plans to boost sortie rate', *Flight International*, vol. 128, no. 3969, 20 July 1985, p. 11.
54. *Aviation Week and Space Technology*, vol. 114, no. 11, 16 March 1982, p. 53.
55. *Flight International*, vol. 125, no. 3910, 14 April 1984, pp. 998-9.
56. Robert Lucas Fischer, *Defending the Central Front: the Balance of Forces*, Adelphi Paper 127 (International Institute for Strategic Studies, London, 1976), p. 33.
57. Neville Brown, *Strategic Mobility* (Chatto and Windus, London, 1963), p. 175.
58. 'Air Defence Mythology', *Air Clues*, vol. 37, no. 6, June (1983), p. 205.
59. For a senior RAF interpretation of these several campaigns see M. J. Armitage and R. A. Mason, *Air Power in the Nuclear Age, 1945-82*

(Macmillan, London, 1983), chs 5 and 8.

60. Max Hastings, *Bomber Command* (Michael Joseph, London, 1980), p. 203.

61. A. H. Quanbeck and A. L. Wood, *Modernising the Strategic Bomber Force* (Brookings Institution, Washington, 1976), p. 65.

62. Basil Collier, *The Battle of the V-Weapons, 1944-45* (Hodder and Stoughton, London, 1964), pp. 151, 165.

63. For a fuller review of the latest evidence from situations of actual combat see Robert K. Harkavy and Stephanie Neumann (eds), *The Lessons of Recent Wars: Approaches and Case Studies, The Lessons of Recent Wars: Comparative Dimensions* (D. C. Heath, Lexington, 1985).

2 CONTEMPORARY AIR DOCTRINES

The West

In a sense, the two streams of doctrinal thought identified in the last chapter were brought together by Germany's blitzkrieg theorists. Their assignation of bombers to the paralysis of key cities and other nodal points to rearward complemented the prime blitzkrieg tactic of disruptive armoured thrusts. But then after 1945 the formulation of air doctrine everywhere became episodic and superficial. The style remained too exhortatory. Too much was made too abstractly of 'ubiquity' and 'flexibility', the virtues classically seen as possessed by air arms in a very special way. Correspondingly, little note was taken of the variations, topographical and otherwise, between different theatres. Little attention was paid to the relevant aspects of human psychology, nor was any attempt made to incorporate the definitive work done since before 1939 on the ethical and legal constraints on the exercise of air power.[1]

For a long time, few people felt this hiatus mattered much. But during these last 15 years dissatisfaction with the said state of affairs has gradually welled up. The progress of historical research about past air campaigns has contributed to this. So has some disconcerting experience of contemporary war. Unfortunately, however, few countries can boast enough political leverage, aeronautical experience and staff development to contribute much to the evolution of new doctrine. Perhaps the only four inside the Atlantic Alliance are France, the Federal Republic, Britain and the USA.

Within the last two, classical notions about the strategic use of air power remain pervasive. That apart, however, there are wide divergences between these several nations regarding operational aims and means. Also, intra-national differences of professional opinion persist, even about the scope of general doctrine. Thus the United States Air Force (USAF) is less keen on conceptual elaboration than it believes the US Army to be. And nowadays, of course, there is a broad undertow of more fundamental scepticism about the future of manned warplanes. None the less, in 1976 NATO at last felt able to publish an encapsulation of its philosophy of tactical air warfare.[2]

A cardinal theme in this presentation (which is to be updated early in 1986) is a need for air power so to exploit its inherent flexibility as 'to adapt to any scale or intensity of conflict envisaged' (NATO, 1976, para. 306). Clearly this stipulation acknowledges the extent to which — in all situations short of a heavy nuclear exchange — political constraints will continue to supervene. However, we are told that, in respect of reconnaissance, these must always be kept to 'an absolute minimum', perhaps through the use of Remotely Piloted Vehicles (RPVs) instead of manned aircraft for missions close to or in adversary air space. But with this cavalier caveat, NATO may just have skirted round a serious issue of principle. For whether or not RPVs alone are used, something approaching an insistence on anticipatory and intrusive surveillance could be read as destabilising. After all, there is bound to be a double-edged relationship between reconnaissance and sudden attack. The former may be an insurance against the latter, yet it may as well be a preparation for it. History is only too rich in relevant materials.

Later on, the text sees an 'extensive and accurate knowledge of enemy electronic systems' (NATO, 1976, para. 805) as constituting the very foundation of Electronic Warfare (EW). Yet in peacetime, attuning Electronic Counter Measures (ECM) to particular mixes of enemy equipment must usually be based on incomplete intelligence. Meanwhile, it is 'extremely important' that every allied asset is secure against espionage.

The key to successful ECM is no betrayal of one's own dispositions: 'ECM should be employed only when it appears almost certain that detection has taken place or as part of a coordinated plan to create confusion within an enemy's defensive system' (para. 808). Among the relevant techniques are 'electronic noise or deception jamming, the employment of mechanical devices such as chaff or infra-red flares or the launching of aerodynamic decoys' (para. 810). A further possibility is to abrade Electronic Warfare (EW) networks, perhaps by means of missiles specifically designed to home on radar sites. Indeed, the mere presence of such a capability may cause an enemy to switch off.

Earlier, one has been reminded that tactical aviation is 'particularly responsive to one of the principles of war, that of concentration of force' (para. 203). However, this blessing is not unmixed. 'Over-centralization or over-delegation of command and control of air resources can result in either inflexibility or inefficiency of their employment' (para. 203). Likewise, in surface-to-air defence, 'Control should be exercised at the highest practicable level. However, it may be delegated to the surface-to-air weapons units involved if circumstances so dictate' (para. 506). At sea, indeed, the delegation of tactical co-ordination to the 'lowest possible level' (para. 511) emerges as a cardinal principle. Correspondingly, anti-aircraft defence tends to demand some dispersion of naval units although such ancillary considerations as emission policy may indicate 'a degree of concentration' (para. 511). So, the text could have added, may surface and submarine threats.

The structure of an air defence system (meaning, above all, the balance between air and surface weapons) depends 'on the geographical characteristics of the area to be protected . . . and the inter-relationship with offensive air operations' (para. 506). While a surface-to-air weapon may best ensure a rapid reaction, a manned aircraft 'is particularly valuable for the identification of a suspected hostile aircraft' (para. 506). Interception from base, whether to inspect or destroy, may be ineffectual without an adequate radar ambience. In any case, Combat Air Patrol (CAP) missions will often — it is claimed — be a better answer, especially against low-flying intruders. Alas, not every commentator would be content with so bald an endorsement of CAP (see Chapter 7).

The interrelationship of all this with offensive air operations is here deemed to have two main aspects. First, there is the need for 'exacting centralised co-ordination to prevent mutual interference' (para. 506), a particular hazard being the shooting down of friends. Two alternatives commend themselves: positive management or procedural management. The former relies heavily on automatic transponders for Identification, Friend or Foe (IFF). The latter 'includes techniques such as the segmentation of airspace by volume and time and/or the use of weapons control orders' (para. 308). We are reminded that terrain and weather ought to determine considerably how this balance is struck.

The second aspect identified is that, in contests for air supremacy, the best means of defence has often been attack, either in the form of fighter sweeps or as strikes against airfields. However, the continuation or escalation of a counter-air offensive should rest upon 'careful consideration' of the military response 'such action may provoke' (para. 408). Indeed, such operations 'may be restricted by political considerations during the initial and possibly during subsequent stages of conflict (para. 408). Nevertheless, defensive and offensive 'counter-air operations may demand the highest priority of all air operations whenever enemy air power presents a significant threat (para. 404).

On the other hand, 'In some situations, it may be necessary to mount sustained or repetitive air interdiction operations for long periods' (para. 411), interdiction here being formally defined as aerial activity intended to destroy, neutralise or delay enemy military units and supplies before they 'can be brought to bear effectively versus friendly forces'. In other words, they are conducted 'at such distances from friendly forces that detailed integration of each air mission is not required' (para. 409–10).

Among the precepts that should shape an air interdiction campaign are these. Surface operations should concurrently be intensified so as 'to accelerate the consumption of the enemy's resources' (para. 411). Although 'immediate air interdiction requirements may be satisfied by using aircraft on ground alert or by re-targetting airborne aircraft' (para. 411), the majority of such missions ought to be preplanned. Once again, political constraints — above all, any acceptance of sanctuary areas — have to be borne in mind.

Armed reconnaissance is here defined as 'an air mission flown with the primary purpose of locating and attacking targets of opportunity'. The dexterity connoted by this role casting may be enhanced by thorough pre-flight planning, the ability of the control facility to redirect in flight, a judicious choice of ordnance, and the 'experience of the aircrew' (para. 411).

With reconnaissance in general, the requirements 'may exceed the capability of the resources available'. Therefore the relevant assets 'are usually controlled from a high level' (para. 608). Also, data about mobile targets 'should be transmitted in real or near-real time' (para. 608): that is to say, as close to instantly as possible.

Close Air Support (CAS) is another classic requirement. Augmentation, in this way, of:

> the firepower of surface forces . . . may contribute decisively during break-through or counter-attack operations by friendly forces, and during enemy assaults and surprise attacks; and is particularly important to offset shortages of surface firepower during the critical landing stages of airborne, airmobile and amphibious operations (para. 415).

A standard variation on this theme is 'column cover': affording specific protection to convoys of trucks, armoured columns, trains and helicopters.

A sphere in which centralised allocation is felt to be especially needful is tactical air transport. One consideration is that the prospective tasks are diverse in character: airborne assault; flare and leaflet drops; ambulance missions; and so on. Another is that some people always feel the urge to take to the air, even when cheaper modes of transit will do. A third is mission overlap between tactical and strategic squadrons. Thus, the performance 'of some tactical air transport aircraft permits their use in the strategic role' (para. 707). Contrariwise, the versatility 'of some strategic air transport aircraft permits them to be . . . tactically loaded outside a theatre of operations, and flown — possibly over relatively long distances — directly into a combat zone' (para. 706).

Among the guidelines laid down for NATO's tactical air arm as a whole are the following. Aerial refuelling may make an important contribution, either by increasing the payloads that planes may carry or else by extending their flight times. Interoperability — meaning, in particular, being able to operate out of airfields controlled by other nations — may underpin crucially the versatility and flexibility inherent in tactical air power. Civilian communications 'may be used for tactical air operations provided there is sufficient capacity to ensure acceptable reaction time' (para. 204).

Meanwhile, stress is placed on the need to sustain operations, intensively and continuously, 'over prolonged periods of time' (para. 204). However, there is implicit acknowledgement of the dilemma presented by many NATO squadrons having both a conventional and a tactical nuclear role: the conventional-phase attrition 'of the forces available for the nuclear option must not be allowed to reach the point where they are no longer credible' (para. 204). Meanwhile, surprise 'remains one of the basic principles in the employment of air power' (para. 203). But the other side of this coin is that aircraft may well be 'the first to offer effective resistance' (para. 204) to any surprise attack.

Yet perhaps the most crucial doctrinal choice of all is whether to go first for complete air supremacy, ignoring all other objects until its achievement. Though deemed a desirable goal, it is acknowledged to be one which may not be feasible or economical to attain. Where absolute supremacy is unattainable, the aim should be to seek a degree 'limited in both time and space' (para. 203).

Manifestly, this overview is wanting in certain respects. It skates too lightly over the problem of airfield vulnerability. It does not explore the human factor. It nowhere sorts its tactical priorities. Still, a question always to pose about such an alliance presentation is how far it may conceal doctrinal differences between respective national staffs. Customarily, the chief divergences of outlook have been between the USA and the West Europeans. For the latter tend not to share at all fully the former's high expectations of exotic technology. As much can be discerned in the decidedly mixed European reaction when the USA first began to advocate NATO's adoption of the Boeing E-3A: the highly specialised Airborne Warning and Control System (AWACS).

A decade ago, the underlying trans-Atlantic contrast in perceptions about 'the future air battle' as such was most evident *vis-à-vis* the projection of air striking power across hostile skies. Possessed as it was of unrivalled electronic virtuosity, the USAF sought (even in Central Europe) to battle in at medium altitude, thereby gaining range and flexibility but forfeiting the natural protection from air defence afforded by restricted horizons at low altitude. What this approach depended on, in fact, was a complex assortment of electronic warfare, escort and ground-attack aircraft; and, since a sizeable proportion of the ground-attack machines would have to be tasked to the suppression of anti-aircraft artillery, only a minority of such an aerial armada could strike at the main surface objectives. Originally, the tight tactical control that a foray of this kind requires was seen as a prime justification for AWACS.

But by 1975 the USAF — together with the Marine Corps air wings — were waxing keener on flying in low. Soon it was even being suggested that previous reluctance on this score had partly stemmed from the 'semantic confusion of terming the dead man's altitudes of 500 to 1500 feet as low'.[3] For some years, of course, their opposite numbers in Western Europe had been persuaded that it was almost always better to go in really low. Nor was this only due to the Europeans' sense of electronic inadequacy. A more positive consideration has been their familiarity with European weather; and with the terrain that must be 'hugged' in low-level flight. A third factor is an emphasis on individual strike aircraft preserving their autonomy in navigation as well as for electronic warfare and tactical evasion;[4] and a fourth may be a greater predilection for attack with area munitions as opposed to forms with precision-guidance.

Underlying all of which is the traditional European and, above all, Clausewitzian scepticism about ever being able to control war with any exactitude. The same goes for the contrast which was (and probably still is) to be observed in the sphere of air-to-air defence. The USAF has been more interested either in highly systematic interceptions by means of missiles of intruders still some distance off or else in dogfights managed sufficiently tightly to continue even within the 'envelopes' of air space covered by SAMs. Conversely, its European partners have thought more in terms of a mêlée, this in between rather than over the successive belts of SAM defence.

Not that there have been no such differences between the nations or armed services indigenous to Western Europe. In the early 1970s the Royal Swedish Air Force (RSAF) led a doctrinal shift towards what was termed Battlefield Interdiction, attacks on the 'second echelon' of enemy forces as they prepared to follow up an offensive. The shift in question was impelled considerably by Alva Myrdal and other leading members of the Social Democratic governing party, informed as they then were by the conviction that 'second echelon' was a more legitimate type of target than, say, Leningrad and Murmansk. Such reasoning, coupled with Leftist resentment at all the money and skill absorbed by Defence, helped bring about a reduction in the procurement of Viggen strike-fighters from the 800 originally envisaged to the 240 in the current programme.[5]

As the notion spread into NATO Europe, it led to a curious inversion. The British, along with the French, came to see Battlefield Interdiction as very generally meaning strikes less than 100 kilometres in depth, whereas some in the Luftwaffe staff set the normal limit at 300. Previously, however, the RAF had been much more inclined than the Luftwaffe to strike deep. After all, this was redolent of its tradition of strategic bombing. Moreover, the extra range planes thus required also helped further a British aspiration to stay globally mobile. But the Luftwaffe's heritage was more to do with the direct support of the ground forces. Besides which, there had been an apprehension in Bonn as well as elsewhere that the USSR would read as provocative too great a German concern with extended range operations. Witness an initial reluctance to have the Luftwaffe supplied with Phantoms. With the burgeoning through the early 1970s of a kind of detente, however, that particular consideration was to wane.

Late in 1982, 'second echelon targetting' was broadly endorsed by Supreme Headquarters Allied Powers Europe (SHAPE). Soon, the Supreme Commander, General Bernard Rogers, averred that his command could:

> achieve greater depth in its defence by extending its conventional fires well beyond the Forward Edge of the Battle Area (FEBA), that is, by interdicting follow-on forces throughout the enemy's rear to prevent them from reaching and reinforcing the forward battle.

He went on to insist that 'Our nuclear threat compels the Warsaw Pact to echelon its forces as well as raising the risk of massing forces for penetration.'[6]

After which, the further promotion of this idea was to come very much from the United States where definitive work was being done on the use of Emergent Technology (ET) to focus on the presumptive 'second echelon' as part of the 'unified air and ground operations' envisaged in a new Air-Land Battle doctrine.[7] This formulation originated in, and was soon to be much elaborated by, the US Army.[8] It was to be accepted by the USAF in 1983. Yet all that these several departures represent is the mere beginning of what will have to be a very wide-ranging debate about the future application of air power. Witness one admonitory assessment of the legal aspects of interdiction targeting.[9]

This debate will resonate with the contemporary expressions of various traditional themes. One of them is that several of air power's special characteristics (capital intensity, technological sophistication, strategic mobility and tactical flexibility) make it singularly suitable as a means whereby Western nations complement, perhaps at short notice, the efforts at deterrence or defence being made by smaller and/or less developed states. Thus in 1978, France's then Chief of Air Staff averred that his service would remain the best instrument of 'overseas action strategy'.[10]

Meanwhile, Israel has made a number of distinctive contributions to contemporary thinking on air power. After the 1967 war, it adopted the practice of regular aerial retaliation against insurgent bases round its borders, a stratagem soon to be emulated by South Africa. When more regular military threats build up, the Israeli air arm serves as a cover for army mobilisation: its peacetime manning levels being over 75 per cent of its wartime ones as against 35 in the ground forces. Also, it is always seen as the prime instrument for strategic attack on economic infrastructures.[11] Yet now, in Israel as elsewhere, there is more controversy than before over the tactical philosophy of air power, including matters so basic as how far the air force should subordinate itself to the immediate needs of border defence. To some extent, the debate involves conflicting judgements about the political worth in the longer term of such long-range aerial ripostes as that against the PLO in Tunis in 1985. It also involves some about the part strategic pre-emption may play.

Soviet Perceptions

Much of what might be said about Soviet thinking on air warfare effectively applies to that on all modes of armed force. This is because (a) this thought is expressive of wider and deeper themes in the USSR's military and political culture; and (b) the role of the air arm has interwoven with those of the surface forces more completely, historically speaking, than has mainly been the case in the West.

Salient among the themes just alluded to is a continentalist and, more specifically, a Clausewitzian scepticism about controlling warlike involutions with any degree of exactitude.[12] The usual response is to centralise the command function very highly but to insist that the focal point of the network be a human being rather than a computer. This philosophy was well expressed in 1974 by the then commander of PVO-Strany, the home air defence service described below (p. 45). He observed that:

> control over units and subunits in modern conditions makes greater demands on the training of commanders and staffs. To carry out control firmly and confidently, uninterruptedly and with operation efficiency, the commander must have a clear conception of the peculiarities of modern air combat and a good knowledge of the combat capabilities of each pilot, squadron and unit as a whole.[13]

In other words, the human factor is basic.

By the same token, Soviet notions about EW give pride of place to human intervention. Various instances can be found in, for example, *Principles of Electronic Jamming and Reconnaissance*, the first Soviet work in the open literature to expound systematically the science of electronic combat. Thus it notes that when 'overcoming enemy aircraft, there is great significance in the selection of the moment' to commence jamming.[14]

The judgement just quoted about centralised command was introduced by an observation about what 'combat operations in Vietnam' had shown apropos the complexity of offensive operations by modern air armadas. Nor is such gleaning of historical experience at all untypical of Soviet writing on contemporary air power. Account is often taken of that recent conflict in South-East Asia and also of those in the Middle East, especially as evaluated in Western journals. Still, particular attention is always paid to campaigns in which the Soviet air force has taken part. The Great Patriotic War (1941–5) remains by far the most celebrated. Others include the civil war in Spain 1936–9 and the extensive fighting against Japan's crack Kwantung army along the Manchurian border at much the same time.

Some six years later, Manchuria was to be the scene of the final Soviet contribution to World War II: the shattering of an already demoralised Kwantung army in a ten-day blitzkrieg characterised by advances of up to 100 miles a day. Professor John Erickson is among those who contend that this episode 'closely approaches in style and scope what the Soviet command presently envisages in the way of high-speed ground operations . . . a much more realistic "model" than the majority of operations in the European theatre during the period 1941–5'.[15]

Looked at from another angle, however, Manchuria in August 1945 was the culmination of a four-year swing by the Red Air Force from pronounced inferiority to marked supremacy. Associated with this were several more defined trends and patterns. Among them were an increase in the average number of sorties per aircraft lost from 32 in 1941 to 165 in 1945; the regular use of Combat Air Patrols (CAP) of 50 circuiting aircraft in pursuance of Close Air Support; and a growing emphasis on mass concentrations at critical times and places. Some 35 per cent of all operational sorties were devoted to air supremacy, 45 to attacks on forward surface targets, 12 to what the West calls 'deep interdiction', and 12 to reconnaissance.[16] Before the commencement of a ground offensive, the aerial surveillance was comprehensive. Afterwards, it was chiefly carried out around the main axes of advance.

What is not informed by experience against the Third Reich, however, is the intense interest nowadays taken in the helicopter. To an extent, this stems from a growing awareness in naval circles of the merit of having air support in a form that can readily be shipborne. More importantly, a conviction has burgeoned that, in the modern land battle, helicopters may be a most pertinent expression of mass and flexibility, not least in the context of a dynamic ground offensive or against a background of residual radioactivity. Among the tasks identified for these machines is the engagement of opposing rotorcraft and maybe monoplanes. Thus 'if a helicopter with a landing force aboard has perfectly organised small-arms fire, it can destroy air targets at a distance of up to 1,500 to 2,000 metres in a vertical target sector of 45 to 60 degrees'.[17]

A related stratagem may be that, in order:

> to intercept low altitude targets, use should be made of fighters based on forward airfields. It is presumed that in the future the best combat means of coping with these problems will be Vertical/Short Take-Off and Landing fighters.[18]

Shades here, too, of a tactic beloved in the Great Patriotic War: 'The fighters also resorted to ambush tactics, small groups of fighters being placed on small field aerodromes along the probable routes of approach of enemy reconnaissance and bomber planes.'[19]

Not that ambushes of aerial intruders were conducted by fighters alone. On the contrary, anti-aircraft guns were not infrequently moved around to catch enemy pilots unawares. This tactic continues to be employed; and has obvious paraliels, conceptually speaking, with what may be a current trend towards greater mobility in the development of Anti-Ballistic Missile defences.

Attention is regularly drawn to the general need always to hold surface-to-air umbrellas over troops engaged in mobile war or in passage thereto. One method is the alternated movement of the specialist anti-aircraft vehicles: meaning, in particular, the ZSU-23-4 tracked four-barrel cannon. Another is having trained specialists in aircraft recognition assigned at platoon or company level. A third is recourse to blankets of small arms fire against both rotorcraft and monoplanes.

Evidently, too, organic air defence (including, perhaps, earmarked fighter squadrons) has been part and parcel of the development, since about 1981, of the concept of the Operational

Manoeuvre Group (OMG). Such a formation (which is at least a reinforced division in size, and sometimes considerably larger) may wait one or two hours' march behind the Forward Line of Own Troops (FLOT)[20] ready to proceed through a break already created in enemy lines in order to exploit this to the full. According to a Polish article, an OMG will possess and control not only a considerable quotient of anti-aircraft defence but also some aerial support. The latter may operate from airheads the OMG has seized in enemy rear areas.[21]

A heavy emphasis on surface-to-air cover throughout the Warsaw Pact's military structure owes something to the reality that a measure of technological inferiority can be accommodated more readily in such weapons systems than with warplanes (see Chapter 6). But it also relates to the strong Russian artillery tradition. At Borodino, the Russians had just over five guns per 1,000 men as against Napoleon's just over three.[22] Similarly, in 1914, the Czarist artillery corps was stronger than its German counterpart in terms of numbers of guns. Its inferiority lay in the quality of the ordnance; the weakness of the logistic support; and a disposition to be fortress-bound.[23] In modern times, its strongest legacy has been anti-aircraft defence.

Evidently, too, the predilection for fortress deployment has been one manifestation of a dominant concern with the integrity of the national borders, both landward and seaward. An awesome expression of this along the shores of European Russia was the prolonged defence, in the wake of the Nazi blitzkrieg, of successive coastal enclaves — Odessa, Sevastopol, Khanko and so on. Here was expressed, too, an historical concern with naval power, a concern which has been remarkable, given the land- and ice-locked situation of the North Eurasian polity in question.

However, to mention this is to highlight how various are the ways in which traditional doctrine, tactical and strategic, may be reinterpreted under modern conditions. Thus the development of on-board aviation has lately been a prime indication that the Soviet navy has extended beyond coastline protection, be that inshore or further out, and towards global deployment. The first 20,000 ton 'anti-submarine cruiser' (bearing 18 helicopters) was commissioned into the Mediterranean fleet in 1968. Seven years later, the first of the Kiev class of 40,000 ton carriers, bearing a mix of helicopters and Vertical Take-Off and Landing (VTOL) monoplanes passed through the Dardanelles. Now 70,000 ton carriers (probably to be fitted with heavy-duty launching catapults) are under construction. Yet all this is at a time when many in the West are waxing sceptical about the future of the large fleet carrier (see Chapter 10).

Such divergence invites the further question of just how futuristic any Muscovite rethinking of land-air doctrine is likely to be. What does the Soviet concern with the central involvement of the human operator connote for the manned warplane? Is it expected to hold its own against the gun and the missile? If so, is this essentially within the particular context of sudden and massive offensive action, an option conspicuously stressed in Soviet military literature since 1970? Correspondingly, must we accept that 'A Soviet non-nuclear attack on NATO would be initiated by a massive independent air operation?'[24]

Let us make mention of just two of the indications that this could be so. The first is that Soviet commentaries on the Great Patriotic War tell how, until general air supremacy was finally secured in the middle of 1943, much effort had to be directed towards counter-airfield operations. Thus one survey published in Moscow in 1972 records that, between the April and the June of that wartime year, a counter-air campaign accounted for 1,000 German aircraft, over 80 per cent through their destruction on the ground.[25] The only rider to add is that these days (a) all the elements in the adversary's air defence would concurrently be attacked; and (b) surface elements would contribute as actively as possible to this onslaught.[26]

The second indication has taken the form of a big, though perhaps arrested, reorganisation of the aerial command structure. Presumably in order to have more assets available for theatre conflict, the reintegration was begun in 1981 of the Soviet air force proper and *Protivo-Vozdushnaya Oborona Strany* (PVO-Strany).

The latter is the elite force of no fewer than 550,000 men and women (on current NATO estimates) assigned exclusively to the air defence of Mother Russia. Early in 1941, PVO-Strany was accorded direct representation on the high command; and in 1948, it was made a separate service.[27] My own reading of the current indications[28] is that it is now destined to revert to roughly its 1941 degree of autonomy. But now its role will be more that of strategic mobility and less the defence of the homeland as such. The USSR seems to be easing itself away from the fortress tradition. Nor is the US Strategic Defense Initiative for ballistic missile interception in Space likely to do more than interrupt this trend.

46 Contemporary Air Doctrines

Still more worrisome for the air arms of the West is a persuasive reinterpretation of Soviet land doctrine. It is far from certain that Soviet and Soviet-inspired armies would even now deploy in the two echelons that contemporary air theorists in the West have lately been assuming. Thus until 1943, the divisions, regiments and battalions of the Red Army

> had always attacked in two echelons, because this was what was prescribed in the regulations. However, the two-echelon deployment was soon seen to be a mistake. This was because the German defenses at that time did not consist of deeply echeloned lines of well-prepared fortifications but of scattered fortified strongpoints and defended areas. Furthermore, the Soviet forces at that time were not numerically superior to the Germans in men and equipment.
>
> To remedy this, the Stavka ordered the Red Army to adopt the single-echelon formation as the standard mode of deployment for the attack and that order remained operative until circumstances changed later in the war.[29]

Indeed, two of the three fronts that took part in that celebrated conquest of Manchuria 'decided to deploy their armies in only a single echelon . . . The third . . . was the Transbaikal Front; but in fact its two-echelon deployment was not so much of an exception as might appear'.[30] It was simply a response to the constraints and demands of geography.

So today, a single echelon is likely to be used on an extended offensive front, wherever geography permits.[31] The only caveat to enter is in respect of the Operational Manoeuvre Groups, the innovation mentioned above. These are expected very generally to be poised 30 to 50 kilometres back, with each spread over an area of 200 to 400 square kilometres. Yet even a number of OMGs deployed in this fashion could hardly proffer the density or the lateral continuity of a truly discrete second echelon. Nor would their attrition necessarily affect the course of the main ground battle. But assisting the surface arms must always be the supreme aim and object of tactical air activity.

References

1. J. M. Spaight, *Air Power and War Rights*, 3rd edn (Longmans and Green, London, 1949) is a celebrated example.
2. ATP-33A, *NATO Tactical Air Doctrine* (NATO, Brussels, 1976).
3. Steve L. Canby, *The Contribution of Airpower in Countering a Blitz: European Perceptions* (Technology Service Corporation, Silver Spring, 1977), p. 3.
4. *Flight International*, vol. 119, no. 3742, 24 January 1981, p. 224.
5. I. Dörfer, 'System 37 Viggen: Science, Technology and the Domestication of Glory' in Frank B. Horton, A. C. Rogerson and B. L. Warner (eds), *Comparative Defense Policy* (Johns Hopkins University Press, Baltimore, 1974), pp. 465–79.
6. 'Sword and Shield — Ace Attack of Warsaw Pact Follow-On Forces', *NATO's Sixteen Nations*, vol. 28, no. 1, February–March (1983), pp. 16–26.
7. Field Manual 100–5, *Operations* (Department of the Army, Washington, 1982), p. 7.1.
8. Major General John W. Woodmansee, 'Blitzkrieg and the Airland Battle', *Military Review*, vol. LXIV, no. 8, August (1984), pp. 1–39.
9. Lt. Col. J. P. Tomas, 'Legal Implications of Targeting for the Deep Attack', *Military Review*, vol. LXIV, no. 9, September (1984), pp. 70–6.
10. 'L'Armée de l'Air. Problems and Prospects', *Aviation and Marine International*, 74, June (1980), pp. 49–52.
11. Major General Eliezer Tal, 'Israel's Defense Doctrine: Background and Dynamics', *Military Review*, vol. LVIII, no. 3, March (1978), pp. 22–37.
12. Neville Brown, *European Security, 1972–80* (Royal United Services Institute for Defence Studies, London, 1972), p. 57.
13. A. Borovykh, 'Pilots Combat Readiness', *Soviet Military Review*, 1, January (1974), pp. 15–17.
14. USAF edition in translation (USAF Foreign Technology Division, Washington, 1969), p. 460.
15. John Erickson, *Soviet Military Power* (RUSI, London, 1971), p. 73.
16. Ray Wagner (ed.) and Leland Fetzer (trans.), *The Soviet Air Force in World War Two* (David and Charles, Newton Abbot, 1974), ch. 20 *et al.*
17. V. Subbotin, 'Air Defense of a Tactical Landing Force', *Soviet Military Review*, 5, May (1977), pp. 23–4.
18. A. Karikh, 'Fighters versus Low-Flying Targets', *Soviet Military Review*, 12, December (1974), pp. 16–19.
19. Ye Veraksa, 'Fighter Tactics', *Soviet Military Review*, 3, March (1978), pp. 35–7.
20. 'Soviet Operational Manoeuvre Group: A Closer Look', *International Defense Review*, vol. 16, no. 6 (1983), pp. 769–76.
21. John G. Hines and Phillip A. Peterson, 'The Warsaw Pact Strategic Offensive: The OMG in Context', *International Defense Review*, vol. 16, no. 10, 1983, pp. 1391–5.
22. D. G. Chandler, *The Campaigns of Napoleon* (Weidenfeld and Nicolson, London, 1966), ch. 70.
23. Norman Stone, *The Eastern Front, 1914–17* (Hodder and Stoughton, London, 1975), pp. 23–8.
24. Phillip A. Petersen, *Soviet Air Power and the Pursuit of New Military Options* (US Government Printing Office, Washington DC, 1978), Vol. 3, p. 53.
25. P. S. Kutakhov in USAF (trans.), *Selected Soviet Military Writings 1970–75* (US Government Printing Office, Washington, 1977), pp. 240–50.
26. Phillip A. Petersen and Major John R. Clark, 'Soviet Air and Antiair Operations', *Air University Review*, vol. 36, no. 3, March–April (1985), pp. 36–54.

27. Neville Brown, 'Air Power in Central Europe', *World Today*, vol. 39, no. 10, October (1983), pp. 378–84.

28. Chief Marshal of the Air Force Alexander Koldunov, 'Defence of Soviet Air Space', *Soviet Military Review*, no. 4 (1985), pp. 19–22.

29. P. H. Vigor, 'Soviet Army Wave-Attack Philosophy', *International Defense Review*, vol. 12, no. 1, 1979, pp. 6–10.

30. Ibid.

31. Christopher Donnelly, 'The Development of the Soviet Concept of Echeloning', *NATO Review*, no. 6, December (1984), pp. 9–17.

3 THE HUMAN FACTOR

Sustained Offensive Action

Given the primacy of strategic attack in the history of air power in theory as well as practice, it is best to begin an assessment of the human factor with a look at morale-busting as a strategic air objective. But what is hard to find is an example of a sizeable stretch of territory being induced to surrender simply by dint of aerial bombardment. Perhaps the closest approach to such a consummation was afforded by the mid-Mediterranean island of Pantellaria (area, 25 square miles) in June 1943. Just as an Allied assault force was coming ashore on the 11th, the commander of the 12,000-man Italian garrison announced a surrender. This demarche followed some weeks of continual aerial pounding, interlaced with naval bombardments.

More specifically, however, it came soon after the destruction of a distillation plant vital to the supply of water to the military as well as many civilians. So what we cannot speak of even here is a morale collapse, pure and simple.[1] Yet this effect was the very one that the classic theorists of strategic air power had believed would prevail. In a memorandum written in 1923, Lord Trenchard, 'the father of the Royal Air Force', averred that the crux of 'an air war' was the 'pursuit of a relentless offensive by bombing the enemy's country, destroying his sources of supply of aircraft and engines and breaking the morale of his people'.[2] The last-stated aim is seen as the culmination of the first.

Today, we look upon the Anglo–American aerial campaign against Nazi Germany as the supreme expression of this approach. Yet with the outbreak of hostilities in 1939, the British proved reluctant to wage war so totally. Early in 1940, a proposed mining of the Rhine waterway was abandoned in ready deference to French fears of retaliation against Paris.[3] The next January, a directive by the War Cabinet stipulated that 'the sole primary aim' of the bomber offensive should be, for the time being at least, the destruction of Germany's synthetic oil industry. Thus its earnest hope was that the surgical excision of a narrow but key sector of German industry was already within the state of the bombing art.

Twelve months later, however, cumulative evidence of the infeasibility of achieving such finesse with the existing aids to navigation and attack prompted another War Cabinet directive. What this said was that henceforth 'the main and almost sole purpose of bombing was to attack the morale of the industrial workers', this chiefly by means of fire-raids. Eventually, improvements in technology were to enable Bomber Command to launch precise attacks on specific sectors, most notably transportation (see Chapter 1). But it never relinquished the alternative option of saturation attack, despite vociferous opposition from certain churchmen, the broad Left, and military sceptics. Its American partner, the 8th United States Army Air Force (USAAF), similarly played things both ways. Granted, most commentators would allow that it did incline more to selective strikes. Even so, the American contribution to the 'overkilling' of Dresden, in February 1945, was much the same as the British. Nor should one forget the decidedly saturation character of the USAAF offensive against the Japanese homeland.

Since 1945, the intertwined debates about the morality and cost-effectiveness of these campaigns have continued. Thus there is still room for historical dispute as to whether the planning staff of Bomber Command ever came to rely on inducing a collapse of German morale.[4] Those who do believe this purpose became uppermost in its collective mind usually attribute it to a too primitive notion of the part played by the pleasure–pain reaction in human psychology. They aver that in mass society in general, and especially a regime as manipulative as the Nazi, repeated doses of such misery were liable to induce not social disorder but fatalistic resignation.

Helpful though fatalism may be to the agencies of social control, however, it can have a most negative effect on initiative and innovation, not least at top level. In every respect, indeed, the psychological overburden imposed by a constant hostile presence in one's own skies may be most inimical to coherent thoughts and wise decisions. Take, for example, the Fuhrer's insistence that the Messerschmitt 262 jet plane be urgently produced as a 'blitz-bomber' rather than an air defence machine, even though its nuisance value in the former role could not compare with what it might have achieved in the latter. So strange an ordering of priorities was not the product of dispassionate reflection, based on a thorough scan of all available evidence. Rather it was part of a deepening obsession with revenge for its own sake. Witness, too, the decision to target the V-1s exclusively on the London area in 1944, ignoring the ports of embarkation for the Second Front.

Granted that this by no means proves that a prime factor in various bizarre judgements made as the war progressed was the exasperation and anguish of the Nazi leadership over the onslaught against Berlin and other cities, but the possibility that such a reaction did contribute ought nevertheless to be explored. Historical insights apart, however, the psychological impact of air power remains a germane issue. Indeed, it merits a central place in future air doctrine apropos Europe and elsewhere.

The Citadel Image

Meanwhile, psychology exercises other kinds of influence within this sphere. One relates to the reality that, however potent aircraft are as instruments of the offence, they may be badly compromised by the vulnerability of the airfields they operate from. All too often the disposition in air force circles has been not to face up to this. Airfields are seen not as the exposed nub of an intricate and hence fragile panoply of air power but as rock-solid citadels from which an aircrew sorties and to which it returns with assurance. Nor can anyone deny that they do have that aura about them. They are capacious yet with well-defined boundaries. Life within them is close-knit, socially as well as professionally. The scenic background is usually distinctive and its boundaries wide. The historical associations are often rich; and are evocatively symbolised by the centrepieces of the whole ambience, the runways down which planes roar on their way to or from the 'wide blue yonder'.

Yet to focus thus on the runway is also to highlight a pronounced polarity in the structuring of the human side of air power. There is a marked disproportion in numbers between the fliers and their ground support. From it, profound consequences stem. Thus:

> precisely because the Air Force is a vast engineering and logistical organisation with striking parallels to a civilian industrial establishment, its top leadership has had special concern with maintaining institutional values and the heroic fighter spirit. Thus the air force has responded by insisting on the central importance of pilot training for career advancement and for maintaining the ethos of the institution.[5]

52 *The Human Factor*

However, this insistence makes it all the more necessary to insist, too, on a somewhat contrary precept: a continual revaluation of the *modus operandi* of aircrew in relation to an evolving technical environment. Above all, we must explore the interaction between the aviators and the machines they fly.

Aircrew Stress

Among the constraints one notices immediately are the limits on acuity set by the human eye. These include the likelihood that (unless he has undergone thorough corrective training) a pilot staring into a near-featureless skyscape (such as often obtains at altitude) will instinctively focus a mere several metres beyond his windscreen. This hazard apart, however, your average aerial observer would stand a 95 per cent chance of discerning across a distance of 60,000 feet an object which was steadily presenting to him, in full daylight and fair weather, a front of 100 sq. ft and which had luminosity in relation to its background of 0.8 which is 80 per cent of the maximum possible. Under a starlit sky with no moon, on the other hand, the standard distance comes down to perhaps 1,200 ft though this may be partially restored through silhouetting against the starfield.[6]

Still, when a machine is flying fast and low in clear daylight, the most critical factor will be the angular velocity at which the Earth's surface appears to swing past. Fifty degrees a second is the median upper limit for the determination of many particular objects and features. Above 80, almost nothing is perceptible in any definitive sense. Yet even at 300 mph, decidedly slow for a modern monoplane, ground lying 200 ft immediately below will be racing by at about 100 degrees a second.

Underlying each and every specific response of a human being in flight, however, is the adverse effect too rapid an acceleration (positive or negative) may have on him. Normally, this factor is measured in multiples of 'g': one g being the acceleration that a body near the Earth's surface is subject to on account of the gravitational field. As is well known, this is some 32 feet per second per second (Appendix B).

Inevitably, electronic systems also exhibit lag when adjusting to abrupt shifts of position and may, indeed, be peculiarly prone to fail outright then. Yet already such lapses are tending to occur much less readily than when human fragility is thus put to the test. Across a wide range, increasing acceleration causes a progressive decline in mental and physiological performance. Then within this generalised diminution, certain trends (such as a loss of visual acuity coupled with illusory perceptions) are pronounced. An especially critical threshold is, of course, loss of consciousness. Among the variables that determine g-stress are the peak acceleration; the rate at which it builds up; its duration; the clothing and posture adopted by the subject; and his own physiology and attitude.[7]

None the less, it is possible to predict where on the g-scale unconsciousness is likely to occur. Young men of aircrew calibre, though without special training or clothing, can be expected to withstand only 3.6 g, this subject to a standard deviation of 0.25 g. However, this natural limit can usually be pushed up a further 3.0 g by breathing and muscular exercises. Another 1.5 or so can be added by donning what is called a 'g-suit': a comprehensive garment so pressurised as to constrain any gravitational shifts of blood within the body. Finally, a plane can be fitted with reclining seats whereby the differential stress felt by the pilot when the plane is banking is reduced by a proportion equivalent to the cosine of the angle of recline, as measured from the perpendicular. When this angle is 67.5 degrees, say, no less than 2.0 g more tolerance is added.[8] Hopefully then, a strain of up to 10.0 g can be imposed without too severe an impairment of crew performance, provided it is not prolonged or recurrent. Hopefully, too, electronic overrides will be devised to avert or accommodate momentary 'black-outs'.[9]

Even so, the scope for advance is limited. Therefore, negative aircrew response to rapid lateral acceleration is emerging more starkly as a limiting factor in the manoeuvrability, at any rate at low altitude, of tactical warplanes. After all, the fighter aircraft which are coming into service in the 1980s are sufficiently heavily powered and well-structured to pull 6.0 g at high subsonic speeds even at medium heights; and they can be made to pull around 10.0 or 11.0 g closer to the surface.[10] Previously, the effective limits in that latter zone have been in the 7.0 to 9.0 g range.

Cumulative fatigue is a further alarming aspect of the manifold stresses. A British commentary on tactical training puts the point this strongly:

The sheer physical strain of air combat is not easily appreciated by someone who has never known it. The students come down from a half-hour session soaked through with sweat and often so exhausted that it would be dangerous for them to fly another sortie that day. They also find at first that the continuous high-g manoeuvring plays havoc with the semi-circular canals in their ears. They feel all right until they have landed and taxied back to the line, but as soon as they stop and put their heads down to replace the ejector-seat safety pin they find everything spinning round.[11]

Likewise, even fully trained pilots experience acute fatigue as well as nervous stress when flying fast and low (say Mach 0.9 and 200 ft) across undulating terrain. Here the fatigue is much aggravated by the oscillations of several g usually associated with 'contour hugging'.[12]

Training and Recruitment

On this account alone, it is not surprising that the training by, for example, the RAF of pilots qualified to fly fast jet aircraft now costs roughly two million pounds a head and lasts between 70 and 85 weeks. In doing so, it embraces around 300 flying hours, a total closely comparable to that normally required before combat of British and American aircrew in 1944. Fourteen weeks are given over to basic training (flying and otherwise), during which time 64 per cent of the initial intake are discharged from the course as unsuitable. Another 10 per cent are thus lost during the 24 weeks of Advanced Flying Training or the subsequent 17 weeks of Tactical Weapons Training. Finally, 4 per cent fail to qualify in a 15–30 week conversion to front-line aircraft. All in all then, barely a fifth of those accepted for pilot training qualify on high-performance jet aircraft though just over another quarter will, in fact, be doing so on multi-engined planes or else on rotorcraft. The number accepted is only a fifth of the number that apply.[13]

Still, some young men adjudged, either at initial selection or later on, to be of aircrew calibre in the round yet to lack the rather esoteric aptitudes a pilot requires may be able to qualify as air electronic engineers, aircraft engineers or navigators. The British programme for pre-squadron navigator training currently extends across 61 weeks.

Various other air forces sustain training programmes that are very similar in respect of entry, duration and flying time.[14] But under a new and quasi-experimental scheme, the USAF and also the Royal Danish Air Force have a flying course for pilots which is cut to 175 hours. This is partly through an extensive use of static simulators of the aerial environment. However, another factor is that neither of these air arms need to be too preoccupied with the peculiarly difficult interaction of climate and topography encountered in the two Germanies.[15] But in them and all other air arms, pilots and navigators are not considered fully proficient until they have also done several years on a squadron. NATO believes that a pilot in squadron service should optimally do 240 hours flying a year and minimally 180. Not that all NATO air arms register even the latter.[16]

Irrespective of the particular arrangements made, however, the problem of providing enough aircrew at the standard of proficiency now demanded has become acute. Let us note an official Australian apprehension that it may no longer be 'practicable for pilots to be operationally proficient in the whole range of skills required in the air-to-air and air-to-surface roles'.[17] Let us note, too, a warning by a Soviet air force general that:

> The average combat-training level today is inadequate for the pilot to use to the full the potentialities of the aircraft equipment or to carry out combat missions successfully in any situation. Therefore, constant growth of the pilot's combat proficiency should be the main goal in the activity of commanders and political workers.[18]

One Seat or Two?

Regardless of training and indoctrination, however, the workload imposed on the crews of high-performance warplanes is approaching the limits of what is bearable in environments so fraught with fear and physical stress. This is so even with a progressively greater share of monitoring and control being assumed by automated electronics. Evidently, this adverse trend can be broadly explained in terms of continual extensions of all aspects of aircraft performance. But a most particular impetus is coming from the burgeoning of Electronic Warfare (EW).

One manifestation of the resultant anxiety among air staffs has been the insistent quest for better cockpit design.[19] Another is to be seen in all the agonising over whether strike fighters ought not normally to be designed to carry a second crew-member, even though this means an extra thousand pounds or so of weight being incurred directly and perhaps several thousand more through the 'knock-on' effect (see Chapter 4). Naturally, the single-seat solution is best so long as it can be rendered compatible with the incremental burdens imposed by Electronic Warfare 'round-the-clock' combat, laser-designation and so on. This basic preference derives not only from lower cost and more nimbleness in flight but also from the precept that, so long as the pressures do not become too great, both the gathering of information and the interpretation thereof are best integrated inside a single human mind. The argument thus posed closely resembles the one that has lately raged about whether next-generation long-distance airliners should have a third cockpit member. His job would be to assist the pilot himself and the co-pilot with whom he alternates or shares prime responsibility.[20]

Accordingly, some analysts strongly favour single-seat for planes primarily designed for Close Air Support (CAS). Then the provision of a second crew-member is reserved for the more sustained navigation, observation, Electronic Warfare and systems check-out needed for deep-penetration missions in well-defended environments and whatever weather. Conversely, others say that, while the deep-penetration environment may be cumulatively stressful, its peak intensity should be low enough for a single individual to cope. On the other hand, the intense demands often made in the CAS situation can be met only by two men.[21]

The abiding influence of this latter view within the USAF is reflected in the interest lately shown in bringing into service a two-seat version of the A-10 Thunderbolt 2 anti-tank machine. Originally, the Pentagon had been dismissive of the view expressed by Fairchild (this aircraft's parent firm) that a pair of crewmen was highly desirable, especially at night.[22] Meanwhile, the major American LHX programme for a tactical rotorcraft is expected to be based on a single-seat cockpit.[23] Likewise, all the five nations involved in 1985 in the deliberations about a new European Fighter Aircraft (EFA) have been at one in their single-seat preference.

However, one elementary objection to a single-seat cockpit is this. Although it may be designed (as in the F-16) with all-round vision much in mind, a lone pilot will always find it hard to scan enough sky for his own protection. A graphic historical demonstration of the two-seat remedy is afforded by the Ilyushin-2 and -8 Shturmovik ground-attack aircraft that flew so numerously with the Red Air Force in World War II. The cost-effectiveness of this series over the battlefield owed much to a decision, made at a conference of pilots and design-staff in early 1942, to add a second crewman. His prime duties were to keep a look out and operate a rearward-mounted machine-gun.

Not that this is the only conceivable counter to the risk of being intercepted from astern and above. Another lies in formation flying. That is to say, low-level intruders can operate in pairs or fours or *en masse*, maybe with a 'top cover' escort. Besides which, radar-warning receivers are widely installed these days.

On the offensive side of things, the laser designation of a fixed surface target was at first constrained by the need for one crew member (who could rarely be the pilot) to keep locked on, after having discerned the reflection from the target of a laser beam generated by his aircraft or by another one or from the surface. However, designators have now come into service (starting with the F-18 Hornet) which track their target automatically from the moment of 'lock-on'. They will thereby dispense with a continuing human involvement and be the more accurate in consequence.

Adaptation to War

Even so, the upshot of everthing will still be that piloting a high-performance warplane will be so daunting a task that all air arms will find it hard to recruit enough suitable people. But what should be regarded as enough? Here it may be pertinent to cite another commentary on recent Royal Air Force experience, this time apropos crew-plane ratios:

> While the RAF struggles to maintain the ratio recommended by Supreme Allied Commander Europe, it in no case exceeds it. The NATO interim standard ratio is even lower and is far too fragile for anything longer than a three-day war at intensive sortie rates. ... The RAF is some 250 pilots below strength. In the fast jet force, this represents a shortfall of 10 to 13 per cent. Unfortunately, the RAF's view of strength is one well below that required

to sustain a major offensive or defensive effort for more than a week. There are no plans to improve the pilot-aircraft ratio.[24]

Arguably, however, allowance for three notional days is quite sufficient. A dynamic main battle might not last much longer. To which one has to add the sombre logic that the ratio would perforce rise steeply as many aircraft were put out of action, at least for a while, without their duty crewmen being lost. Damage in the air may effect this. So may immobilisation on base.

At all events, to discuss this ratio as a determinant of the resilience in combat of an air arm is surely to accept the following. The ultimate question about the human factor in regard to matters military must always be how it will relate to actual hostilities. It is frequently claimed, most openly in squadron messes, that what air arms do on peacetime exercises has a 90 per cent correspondence with what they would do in wartime. Demonstrably, however, this claim is excessive. Take, for example, the scope for tactical flying presented in peacetime to the NATO air arms in Germany. In a few limited areas, planes are allowed to descend as low as 250 ft. But very generally, the minimum ceiling is 500 ft, which might now be much too high for active operations over a well-defended hostile environment. Rotorcraft are also severely circumscribed, not least because of the noise they make. Often, too, the altitude minima are higher still at night.

Another indicator of constraint is as follows. Scheduled civil air routes in Europe are 16 per cent longer than 'great circle' measurements would lead one to expect. Yet in the USA, the difference is only 4 per cent.[25] A big factor in this geometric anomaly is the distribution of those aerial zones in Europe given over to military convolutions. However, such an interaction is bound to work both ways, at least in a free society. If the military constrict the civil, the converse will also apply. Moreover, it is impossible to use chaff (broadcasted strips of metal foil) or very much else to jam the radio or radar transmissions of the opposing side in a peacetime exercise, not least for fear of disrupting civil telecommunications all over the place. The initial ECM testing of the E-3A Sentry was less rigorous than it might have been. It was said that 'One reason for not carrying out more severe tests is that they would interfere with civilian reception in the US.'[26]

A sphere in which there may sometimes be an acute contrast between inexperience in war and veteran standing is air-to-air combat. The campaign in Vietnam 'demonstrated with stark clarity that a combat pilot who survived his first ten missions was likely to go on surviving the ones that followed'. Still, considerable benefit may have been derived since then, by various Allied air arms, from the Red Flag programme at the USAF base at Nellis in Nevada. 'Here there are six million acres of airspace for simulated combat, over a desert littered with dummy radar and missile sites, guns, trains, truck convoys and other military installations.'[27] Aircrew under training at Nellis have been competing against two 'aggressor' squadrons, flying F-5Es in simulation of MiG-21s. So is this a suitable preparation for taking on MiG-23s during a German winter? Used in conjunction with sophisticated ground simulators (see Chapter 4), it may be so in some measure.

One regard in which a certain continuity is preserved across the peace–war divide is the psychological detachment inherent in the cockpit environment. Usually, this is observed as part of the ethical objection that modern 'antiseptic war' frees the one who fires a weapon from any emotional involvement, and hence of any sense of personal responsibility. As was exclaimed at the 'Winter Soldier' hearings in the USA in 1972, 'The machine functions, the radar blip disappears. No village is destroyed, no humans die. For none existed.'[28]

Yet whether psychic disconnection, as it is sometimes called, necessarily leads to brutal excess more readily than does a face-to-face interaction may be questioned. What does seem credible, none the less, is that psychological distancing is conducive to air squadrons reacting determinedly to the onset of an overland offensive. It thereby makes a singular contribution to the more operational aspects of deterrence.

Yet with the capsulated isolation of the cockpit may go a certain brittleness in the moral fibre of some aviators. Such a trait was evident enough in the suicides in the upper strata of Goering's Luftwaffe as attrition and exhaustion wore it down. Much the same was said soon after World War II by the man who had been the most senior of Britain's tactical air commanders then: 'An air force is, I suppose, by reason of the nature of its work, a highly strung and temperamental body. Any fluctuations in its morale have an immediate effect on the efficiency of its operations.'[29] Even in civil aviation, aircrew optimism and confidence may be an important determining factor. The sudden and big increase in deaths in accidents in 1985 has had something of a global 'chain-reaction' about it.

Where distancing, psychological or otherwise, could never be effected is in the role of Forward Air Controller (FAC). He is the person who directs, from a land vehicle or a helicopter, Close Air Support (CAS) missions against nominated targets. His situation as a direct monitor of ordnance delivery is as parlous as that of the time-honoured artillery observation officer. In view of this and of the difficulties inherent in ground–air co-ordination, an RAF serving officer cautioned in 1972 that 'in the final analysis . . . less than five per cent of the total air support effort' might be directed by a Forward Air Controller.[30] Still, laser designation and other changes may be increasing the likely percentage. In fact, many soldiers have long felt that, if CAS is to remain viable at all, more reliance will have to be placed on the FAC.[31] The danger he faces will have to be seen in relation to the rising lethality of war for all concerned. The importance of his contribution will be judged in relation to the priority accorded to CAS as such (see Chapter 13).

Apropos surface-based anti-aircraft systems, two factors merit a mention. The one concerns that familiar, albeit not invariable, rise over the first few hours or days in the attrition rates registered. As suggested already, this tendency owes much to the fact that those who man the said weapons will then be on a steep learning curve.

The other is that surface-to-air defence against planes transitting fast and low is compromised more than most activities in this sphere by the time taken to transmit perceptions and reactions within the human body. The brain usually takes 0.1 seconds to register a new visual image, then 0.4 to interpret it. Then another 0.5 is required to translate response into muscular action.[32] In the one second thus elapsing, a plane doing 600 mph would cover 880 ft.

In helicopters, the biggest problem is g fluctuations which are not large but are continual and erratic. The biggest challenge is low-level nocturnal manoeuvre.[33] With manned satellites, 'zero g' or weightlessness is conducive to 'space sickness' and to disorientation in orbit as well as to subsequent medical abnormalities.[34] It thereby becomes an argument against prolonged operation by humans in space, unless offset by centrifugal rotation.

Ultimately, the whole debate about military aerospace these days boils down to the question of where various balances are to be struck between such human factors and the electronic. Nor does this just revolve around the manned platform *per se*. It ramifies through every facet of air power, from high policy to logistic support. How far can electronics ease the human burden? How much may it aggravate it?

References

1. Raymond De Belot, *The Struggle for the Mediterranean, 1939–45* (Princeton University Press, Princeton, 1951), pp. 209–10.
2. Quoted in M. J. Armitage and R. A. Mason, *Air Power in the Nuclear Age, 1945–82* (Macmillan, London, 1983), pp. 4–5.
3. *Sunday Telegraph*, 21 March 1971.
4. Noble Frankland, *The Bombing Offensive against Germany* (Faber and Faber, London, 1965), ch. 3.
5. Morris Janowitz and C. C. Moskos, 'Five Years of the All-Volunteer Force', *Armed Forces and Society*, vol. 5, no. 2, Winter (1977), pp. 171–218.
6. *Bio-Astronautics Data Book* (National Astronautics and Space Administration, Washington DC, 1964), pp. 312–13.
7. Ibid., pp. 33–4.
8. D. H. Claister, *Injury*, 9, pp. 191–8.
9. See letter by Geoffrey Moreland in *Aviation Week and Space Technology*, vol. 122, no. 23, 10 June 1985, p. 138.
10. G. Roe, 'The Design of Future Cockpits for High-Performance Fighter Aircraft', *The Aeronautical Journal*, vol. 82, no. 808, April (1978), pp. 159–66.
11. Colin Strong and Duff Hart-Davis, *Fighter Pilot* (Queen Anne Press for BBC, London, 1981), p. 135.
12. See the authoritative letter from F. L. Distefano in *Aviation Week and Space Technology*, vol. 120, no. 16, 16 April 1984, p. 212.
13. *First Report from the House of Commons Defence Committee 1980–81* (HMSO, London, 1981), pp. 9 and 17.
14. 'The Training Scheme: All Things to All Air Forces', *Interavia*, 1, January (1979), pp. 27–30.
15. *First Report from the House of Commons Defence Committee, 1980–81*, p. 47.
16. *International Defense Review*, 5, 1983, p. 572.
17. *Australian Defence* (Australian Government Publishing Office, Canberra, 1976), p. 24.
18. P. Bazanov, *Soviet Military Review*, 8, 1981, p. 18.
19. 'Single-Seat Fighter Cockpits', *International Defense Review*, vol. 18, no. 5, April (1985), pp. 673–92.
20. Bill Sweetman, 'Automatic airliners — ready for take-off?', *New Scientist*, vol. 92, no. 1277, 29 September 1981, pp. 302–6.
21. H. Sackman, *Toward More Effective Use of Expert Opinion: Preliminary Investigation of Participatory Polling for Long-Range Planning* (RAND, Santa Monica, 1976), p. 53.
22. 'Dangerous Midnight', *Defense and Foreign Affairs*, vol. VIII, no. 12, November 1980, pp. 12–43.
23. Mark Broughton, 'The Helicopter Cockpit of the Future', *Military Technology*, no. 6 (1985), pp. 56–8.
24. *Defence*, vol. 13, no. 1/2, January/February (1982), p. 67.
25. *The Economist*, 12 January 1980.

26. J. Burton, 'Under His Wings Shalt Thou Trust', *Defense and Foreign Affairs*, vol. 9, no. 4, April (1981), pp. 9–13.

27. *Flight International*, vol. 125, no. 3918, 16 June 1984, p. 1529.

28. Quoted in Paul Dickson, *The Electronic Battlefield* (Marion Boyars, London, 1976), p. 209.

29. Arthur Tedder, *The Lees Knowles Lectures* (Hodder and Stoughton, London, 1976), p. 209.

30. Wg. Cdr. Duncan Allison, 'Air Support of the Land Battle in the late 1970s', *RUSI Journal*, vol. 117, no. 666, June (1972), pp. 21–9.

31. C. E. Canedy, 'Tac Air: An Army View', *Air Force Magazine*, vol. 61, no. 2, February (1978), pp. 56–7.

32. W. E. Woodson and D. W. Conover, *Human Engineering Guide* (University of California Press, Berkeley, 1964), pp. 6–20.

33. John Rush, 'Cockpit Systems for the Future Battlefield Helicopter', *Defence Helicopter World*, vol. 4, no. 3, September–November (1985), pp. 77–8.

34. 'Will Man Beat the Zero g Barrier?', *Flight International*, vol. 127, no. 3963, 8 June 1985, pp. 51–3.

4 THE IMPACT OF NOVEL TECHNOLOGY

Electronic Warfare

It is all very well saying, as General Beaufre did (see Chapter 1), that at the commencement of any future hostilities, a 'truly enormous experiment' will take place apropos the impact of electronics on combat. Experiment classically involves the testing of a hypothesis. However, we are in no position to hypothesise at all grandly about Electronic Warfare (EW) as such. All else apart, we can never separate it from fire, movement, communication, concealment, disguise and other military activities. Even so, we do need a few basic concepts for the due exploitation in war of this peculiar dimension (see Appendix A).

Recognition of this informs a USAF quest for a set of development packages that will form an Integrated Warfare System (INEWS) for use by American aircraft in the 1990s.[1] Hence, too, an urgent British plea for a NATO electronic plan for 'the physical destruction or neutralisation of enemy emitters', a plan intended to interact with those for field intelligence, artillery and so on within a commander's overall preparations.[2]

Behind all EW guidelines, however, is a fundamental question. How steady and assured, in a real-war situation, will be the flows of data that are the very essence of the electronic revolution? Might electronic coherence soon crumble into electronic chaos? The answer bears crucially on the future structuring of forces, not least those concerned with air warfare.

A vision of chaos is not difficult to conjure. Accidental interference with emissions is always liable to occur. There is often scope for deception with camouflage and dummies. Active Electronic Warfare (i.e., jamming, simulation and attrition) may compromise both communications and surveillance. The increasing pace of war imposes added strains.

From World War II onwards, indeed, there have been instances of command-and-control networks either being reduced to impotence or else surviving only through reversion to quasi-visual surveillance and devolved command. Take, for instance, the introduction of chaff by RAF Bomber Command in the summer nights of

64 *The Impact of Novel Technology*

1943. The Luftwaffe command for the defence of Germany was thereby obliged to revert to a loose and quasi-visual routine for tactical control. Its ground-based radar was now only asked to provide (in conjunction with ground-based visual observers) an indication of the movement of the bomber stream as a whole.

Prominent among the tactics then adopted was *Wilde Sau*: the Luftwaffe fighters converging on a target city lit up by searchlights, flares and burning buildings. As and when these planes could then strike from above against bombers crossing the target area with a layer of cloud beneath them, the stratagem also served as a response to H2S, a new British airborne radar for bombing through thin cloud. Often, too, the light from the surface would diffuse through a layer of altocumulus or whatever enough to enable the *Wilde Sau* fighters to close with assurance on their prey. However, it proved virtually impossible to preserve an altitude divide, below which the anti-aircraft guns were free to engage the bombers and above which the fighters were. So after a sharp increase in German aircraft losses to German artillery shells, *Wilde Sau* was abandoned. Meanwhile, the EW alternation between offence and defence was assuming other modes.

Chaff is still the most familiar form of ECM. It may here be defined as metallic strips each cut to such a length as to resonate at a stipulated frequency. Cumulatively, it produces on radar screens either misleading blips or else a general 'white-out'. To be more exact, a strip of a given length has an appreciable effect within a frequency band extending perhaps 10 per cent (maybe one or two megahertz) either side of the one for which it is ideally tailored. Working to that percentage, a mere kilogram or so of chaff can be shown to present an image equivalent to 30 square metres head-on and solid in all directions and across all frequencies from, say, one to ten gigahertz (see Table 4.1).[3]

With the introduction, in B-52s flying over North Vietnam, of non-manual techniques for the on-board generation of chaff, this stratagem truly became a science: one optimising its results in the light of electronic and meteorological reconnaissance and of calulated rates of chaff descent and retardation. Thus the USAF's ALE-43 cutter-cum-dispenser enables an operator to mass-produce in flight lengths of strip corresponding to whatever points he chooses within the VHF to SHF frequency range. The unit itself weighs 84 kg and its chaff supply another 160 kg.. The dispensing rate is 2 kilogrammes a second. Among many other aircraft that

Table 4.1: The Electromagnetic Spectrum

Band	Frequency	Wavelength
Very low frequency (VLF)	0–30 kc/s	Above 10,000 m
Low frequency	30–300 kc/s	10,000–1,000 m
Medium frequency	300–3,000 kc/s	1,000–100 m
High frequency (HF)	3–30 mc/s	100–10 m
Very high frequency (VHF)	30–300 mc/s	10–1 m
Ultra high frequency (UHF)	300–3,000 mc/s	100–10 cm
Super high frequency (SHF)	3–30 kmc/s	10–1 cm
Extremely high frequency (EHF)	30–300 kmc/s	10–1 mm
Infra-red	—	10,000,000–8,000Å
Visible light	—	8,000–4,000Å
Ultra-violet	—	4,000–10Å
X-rays	—	10–0.01Å
Gamma rays	—	0.01–0.0001Å
Cosmic rays etc.	—	Below 0.0001Å

Notes:
1. kc/s = 1,000 cycles per second; mc/s = 1,000,000 cycles per second; kmc/s = 1,000,000,000 cycles per second.
2. A term now regularly employed to indicate frequency is the hertz, its meaning being 'one cycle per second'. Correspondingly, a kilohertz is a thousand, a megahertz a million and a gigahertz a billion cycles per second.
3. Å = Ångström = one ten-millionth of a millimetre.
4. It is very standard practice to indicate by abbreviation only (HF, VHF etc.) those frequency bands within the range, 100 m to 1 mm. The same goes for VLF.
5. It is also very standard practice, in frequencies up to the EHF, to relate frequency to wavelength on the basis of their multiple (that is, the 'velocity of light') being taken as precisely 300,000,000 metres per sec (mps). To be more exact, it is 299,792,500 mps.
6. The spectral delineations indicated above had become customary in the West by the 1960s. A more recent one is referred to in my Appendix A. For sereral more historical formulations see Lt. Col. Richard E. Fitts, *The Strategy of Electromagnetic Conflict* (Peninsula Publishing, Los Altos, 1980), Appendix A.

carry chaff are certain fighters, usually 10 or 20 kg.

Yet even though chaff can also be used in association with decoys and jamming techniques, some commentators are disposed to play down its utility. They point out that the several seconds it takes fully to 'bloom' could be critical. But more do they stress that — especially in the dense lower atmosphere — retardation together with descent may separate a chaff cloud from the aircraft it is supposed to shield or, at any rate, enable the opposition to distinguish between them by doppler shift (see Appendix A).

However, recourse to air-launched rockets or unmanned vehicles to release chaff above and ahead of intruding aircraft and cruise missiles may appreciably offset the inevitable drifting astern and below. Moreover, the B-52 sorties over North Vietnam in the 'Linebacker' offensive of 1972—73 were conducted within corridors of 'white-out' too continuous to allow doppler discrimination. Nor need descent rates prove unacceptably great in general, three to five feet a second being normal at medium altitudes.

Besides, though chaff can rarely be employed in peacetime exercises (see Chapter 3), its worth has been confirmed in recent air wars. In its absence, the B-52 losses during Linebacker might well have been an order of magnitude (a factor of ten) higher. Aircraft and also naval vessels employed it quite successfully during the 1973 war in the Middle East. And to take a non-lethal contingency, a chaff corridor 200 miles long and persisting several hours more or less screened from NATO the air dimension in the Warsaw Treaty deployment into Czechoslovakia during the 1968 crisis.

A particular attribute of chaff is that it can thwart what might otherwise be two of the most formidable kinds of Electronic Counter Measures (ECM): frequency-agility and monopulse. Waveguide considerations (see Appendix A) inhibit most radars from altering their frequencies by more than a few per cent in either direction, a spectrum which could readily be swept by an ECM observer. Nevertheless, an agile radar operator might keep a jump ahead often enough to get an adequate succession of reliable returns.

If so, the other side may need to resort to chaff. The only alternative is 'barrage' jamming: the simultaneous blocking of a wide range of frequencies. Alas, missiles may be able to home too easily on the source of such indiscriminate noise. Besides which, barrage jamming makes heavy demands on electrical power: a limiting factor in all vehicles, aerial or otherwise. There may also be collateral interference with friendly transmissions.

The principle of the monopulse has long been applied. An example is the tail gun of the B-52. Through the simultaneous employment of several (normally five) antennae, a radar can get a true target bearing from one pulse alone. Therefore, it cannot be spot jammed or spoofed or used to home on. With five antennae, four are set as though at the corners of a square, while the fifth is placed at the centre. The signal is emitted from this central antenna or else from all five together; and the echo is received by all five. The difference in received signal strength at each end of each diagonal affords a measure of the pointing error along that axis.

A concept that, through the very hesitancy of its progress,

continually figures in the dialectic about cohesion and chaos is Identification Friend or Foe (IFF). The principle is that an automatic transponder on board the aircraft or whatever will pick up an interrogatory signal and transmit a pre-arranged reply. In the absence of such a response, the object of the interrogation is presumed to be hostile.

However, this field is one in which a wide gap has long persisted between theory and reality. It was largely as a result of IFF deficiencies that several Israeli and maybe dozens of Arab aircraft were lost on the Golan front in 1973. According to Israeli reports, in fact, no fewer than 20 Iraqi aircraft were destroyed by their own side in a single day.[4] Sometimes, NATO exercises still reveal an 'alarmingly high' incidence of the same aberration.[5]

Similarly, the Vietnam war was engendering debate in USAF and USN circles about a seeming inability to devise IFF equipment that was much more than 50 per cent reliable[6] in combat environments. Hence the standard rule was 'positive visual identification' before opening fire, a viable solution in a situation in which adversary formations were in a decided minority but one that could prove hopelessly constrictive in, say, central Europe.[7] Meanwhile, the IFF systems so far available in NATO Europe are tied to 'operating techniques which could easily cause self- and mutual-jamming in a hostile environment'.[8]

One hazard is sheer saturation. The interrogator part of a contemporary surveillance radar tends to perceive as interleaved the transponder responses from two aircraft lying in roughly the same direction and not more than 3 km away from each other. Hence its chances of eliciting an uncorrupted first reply drops from 90 per cent if there is one aircraft within its beam, to 72 per cent if there are five, and below 1 per cent if 20.[9]

But no less to be reckoned with is adversary interference. Witness the way in which — during the strategic bombing campaign over Nazi Germany — the Luftwaffe anti-aircraft command learned how to activate the transponders of RAF Lancaster bombers over 50 miles away. To which one might add that, even in the benign and cooperative environment of civil air traffic control, a 1 in 25 initial non-response is conceded with the latest transponder links.[10]

Another sign of slow progress is how awkwardly NATO has been groping towards a common IFF programme ever since 1969. Even now it is no more than just possible that a NATO Identification System (with radar and laser modes of interrogation used in parallel) will enter service, on the surface and in the air, around 1995. The prospect is that slim partly because of the amount of money already invested in national systems, some five billion dollars in the case of the USAF.[11] Relevant, too, are Euro-American conceptual differences, some of which relate to the deeper and more philosophical divisions about the very nature of command and control (see Chapter 2). The Europeans prefer to concentrate on the perfection of IFF as narrowly defined. Most Americans want the on-board IFF processors to blend continuously into the process of positive identification a flow of background information about friendly and adversary dispositions.

No matter how that dialectic develops, however, it is likely that NATO operational routines (and, in all probability, those of the Warsaw Pact and other powers) will have to include defined corridors through which — and stipulated times during which — friendly aircraft may transit to and from hostile territory. In any case, it is always desirable to have a synoptic appreciation of the pattern of hostile or friendly air movements. But in the absence of standardised and secure IFF equipment, the alliance may still have to live with something between a 25 and a 60 per cent weakening of its air defence potential.[12] Investment in more weaponry as such might offset this shortfall only marginally.

Moreover, there are no signs of an early IFF breakthrough via a less orthodox route. Granted, automatic voice-recognition systems are even now available which can cope with (a) a 100-word vocabulary being employed by a small number of users whose speech characteristics the computers have been trained to recognise; or (b) a six to twelve-word one when communicating with strangers. However, the accent within the USAF is currently on possible applications to control 'without wires' within individual planes.[13] Surely, it could never provide IFF in anything but a low density environment.

An alternative approach is to recognise the type of machine through an electronic analysis of radar reflections from it. According to a research paper published in 1975, just four discrete frequencies ought to be enough to allow the positive identification of an aircraft type. Meanwhile, British Aerospace has been reporting:

> some success with a millimetre-wave radar system. During field trials, radar returns from targets showed 'extensive supplementary information', and the company considers that the

experimental equipment offers new prospects for all-weather target classification and location.[14]

Laser radar may also be helpful.

Less radical and so more immediately relevant are the electro-optical aids to visual recognition: recourse to telescopic optics and maybe light amplification. Hopefully, these concepts will permit aircraft recognition and tracking even at the ranges achieved by the larger air-to-air missiles. Already first-generation electro-optical devices are fitted to various kinds of fighter, including the Air Defence Version (ADV) of the Tornado.

But even allowing for the potentialities of the other techniques here mentioned, it is clear that the chaff and IFF situations will alone be enough to prevent the electronic revolution consistently affording a full and coherent flow of information throughout a conflict zone. On the other hand, Electronic Warfare is unlikely to undermine comprehensively the opposition's surveillance and control capabilities except as and when it can exploit a large technology gap. Otherwise the pattern will be patchy and erratic on both sides. At certain times, the data flow may be voluminous and definitive. At others, it may all but dry up. Coherence and chaos will alternate, sometimes sharply.

Accordingly, one has to ask whether those addressing the development of air warfare are not sometimes too ready to slide into a state of electronic dependence. Thus in 1950, a part-time Ground Observer Corps was established in the USA in anticipation of intercontinental air war. In 1956, it was to reach a peak strength of 390,000 volunteers and 1,700 'round-the-clock' posts. Yet as early as 1959, it was peremptorily deactivated, less on account of diminished apprehension — as early as that — of air attack *per se* than because of a burgeoning conviction that radar surveillance was the complete answer. Nor did the USSR evince much concern through the 1960s with visual back-up for its home air defence radars despite persisting gaps in their coverage at lower altitudes.[15]

Elsewhere in Europe, however, visual observation from the ground has been less readily discarded. No matter that even shoulder-launched missile launchers may now be fitted with IFF interrogators. Several countries retain a network of visual observers. Take, too, a Soviet army prescription for the protection of a small riverine bridgehead:

Air observation is conducted visually with the use of optical devices . . . The air observers are in a position to spot targets visually at a range of five kilometres . . . there should be at least three trained observers in each platoon . . . and they should be relieved every 30 minutes.[16]

At all events, the prospect of recurrent lapses into electronic chaos must reduce the expected rates of destruction in flight of aircraft or missiles. Added to which is the deliberate stratagem of minimising at the design stage the radar image they reflect back. Admittedly, propellers and rotors are bound to be highly reflective. But otherwise, there is scope. Seen from directly above or below, a large modern warplane may well present an image equivalent to that cast by a perfect reflector some 50 square metres in extent and held perpendicular to the axis of the beam. On the other hand, an image of only 2 or 3 square metres is projected head-on, at least in the absence of squarish air intakes or other angular surfaces. Furthermore, the head-on image from a Remotely-Piloted Vehicle (RPV) or cruise missile may be more like a quarter of a square metre.

In addition, an image may be further reduced (maybe by a factor of five) by means of radar-absorptive paint. Around 1960, much work was done on this in the United States, the paints being divisible into two main categories. With the one, the cancellation of reflected waves was sought by inducing mutual interference. But while such coats could be just millimetres thick, they ceased to be absorptive as soon as the incident frequency shifted from their resonant one.

With the other sort of coating, the reduction sought was across a broad band of frequencies. However, this involved the paint being spread more thickly, perhaps to a depth of two or three centimetres. Moreover, both this kind of paint and the other flaked too easily.

Subsequently, reports about this mode of ECM became so sparse as to suggest that little faith in it endured. Also what interest was still being taken related more to the disorientation of homing missiles than to the grander task of deceiving master radars. But then, in August 1980, the US Department of Defense affirmed the existence of a new Stealth technology designed to 'degrade radar and other sensor detection integral to air defence systems'.[17] Apparently, Stealth (which is based on new combinations of absorbent materials and deflecting shapes) has been integral to the development of cruise missiles and RPVs since the mid-1970s. Now the intention is that, in the 1990s, it will figure in the design of most new aircraft, including any new strategic warplane. Thus the B-1B bomber may project radar images below a square metre head-on while the Advanced Technology Bomber (ATB) provisionally destined to succeed it is intended to manage several times less. The latter will be built very much around the Stealth concept.

Shortly before he retired as US Defense Secretary in 1981, Dr Harold Brown reiterated his opinion that Stealth 'may be the most significant military development of this decade'.[18] Correspondingly, there has lately been a Pentagon claim that the ATB will be 'all but impossible' to track, even when operating at high altitude.[19] But air warfare in general (and perhaps surveillance in particular) is a field in which technological prognoses tend to be too brash, and this may be no exception. After all, a plane approaching at height obliquely presents its broad undersurface to ground-based radar. Then again, a four-fold reduction in a projected image will only halve the maximum range of detection, all other things being equal (see Appendix A).

Besides, other things will not be equal. Radars operating at altitude or at millimetric wavelengths are likely to be harder to deceive with Stealth. Much more important, however, is the way advances in the computer resolution of radar images is far more than outstripping the tendency of those images to diminish in size. Thus a typical assessment of the overall trend has been that, whereas the performance of representative computers improved three-fold between 1965 and 1980, a further 30-fold advance is to be expected by the end of this decade.[20] Nor will this acceleration stop there.

The Future of Aeronautical Science

In the field of aeronautics as such, the development prospects have often been determined by the efficiency and power of the means of propulsion available. Through the turn of the century, however, improvements on that front are likely to be incremental rather than radical. Take nuclear power. The onset of the energy crisis in 1973 did lead to some resurgence of the active interest in nuclear-driven manned aircraft evinced in the mid-1950s. Even so, the needful weight of reactor and shielding still precludes this solution being considered for machines much below 2,000,000 lb all-up-weight,[21] over twice as heavy as a Boeing 747 'jumbo' jet. Besides, anxiety is now too widespread about nuclear energy *per se*, even when employed on terra firma.

Nor do the two chief possibilities as new chemical fuels — namely, methane and hydrogen — hold out great promise. The former (see Chapter 13) is as easy to handle as is kerosene and is readily available within natural gas, the production of which should peak later than will that of petroleum. However, its liquified volume per calorie is 70 per cent more than is that of kerosene even though it weighs 14 per cent less.[22]

Meanwhile, liquid hydrogen takes up three or four times as much room in flight — for a given calorific value — as kerosene does, although its weight is 65 per cent less. Nor is the former as yet at all easy to store or handle or even economic to produce. Nevertheless, one informed view is that it will figure in aviation (and especially in supersonic transports) by the year 2015.[23]

As regards the more evolutionary choices in engine design, the turbine and turboprop are presenting the piston engine with still fiercer competition than before at the lower end of the power spectrum, a trend that has favoured inter alia the stronger emergence of the cruise missile. Further up, development continues of the turbofan: the type of jet engine in which a lot of the air inflow passes along an annular duct between the turbine core and the exterior engine casing. Considerations which appertain to supersonic agility will keep the by-pass ratios low in tactical warplanes. However, those achieved when such engines are installed in larger or more staid machines should rise from a maximum of c.6.0 today to 7.5 by the century's end.

What is more, this trend has aroused enthusiasm for the first rotor in the cycle no longer being cowled at all, for its reverting thereby to being a propeller — albeit one so shaped (maybe with eight blades, thin and well swept back) as to curb the compressibility problem that propellers of the classical kind face as the sound barrier is approached. Nevertheless, noise may still be hard to reduce to levels that are commercially viable;[24] and the commercial applications (essential to low-cost production runs) may be further vitiated by the supposedly reversionary nature of this concept. Also to be reckoned with is the strong radar image even a swept-back propeller will yield. On the other hand, propfans might enable largish aircraft to travel some 25 per cent further unrefuelled in the

1990s than turbofans will allow; and this at speeds of up to Mach 0.8.[25]

Nor is the large turbine without room for further improvement. Inlet temperatures will rise from the 1,600° C representative now to 2,000° C, the level at which the combustion of kerosene is at its most thorough. Likewise, as compressor temperatures are raised from the 650° C now current to 850° C or thereabouts, the specific power generated in the engine core could double. At the same time, however, the amount of air bled off to cool engine hot parts ought to be 'reduced from around 30 per cent today to 10 per cent'.[26]

Evidently, progress along these various lines depends primarily on advances in materials science. While the use of ceramics may almost be confined to engines and blades, other new materials will also be employed in airframes. Titanium and its alloys continue to command attention, a prime concern (outside the USSR, at any rate) being to lower this metal's bulk costs. Titanium's endurance of heat facilitates the raising of air speeds above what otherwise is, even at higher altitudes, an awesome barrier of thermal resistance around Mach 2.8. Witness its extensive use in the airframe of the Lockheed SR-71, the strategic reconnaissance plane which does well over Mach 3.0. Other most positive attributes of this metal are its resistance to corrosion and its fracture toughness.

Obversely, it seems 'that the high-strength steels have just about reached their practical potential in strength'.[27] Nor has much progress lately been made in the application in aerospace of aluminium alloys, a lapse which relates to a drop in this sector's share in aluminium consumption from over 90 per cent in 1944 to about 4 per cent today. Still, aluminium alloys do stand to benefit a great deal from 'powder metallurgy'. Here the cooling rate of a molten mix is accelerated by orders of magnitude. The result can be much improved microstructures coupled with weight economies of perhaps a third.

Meanwhile, the application of the advanced composites is gathering pace. These are two or more (often non-metallic) materials plied together to yield a desired blend of properties. Carbon fibre is among the more firmly established of the non-metallics. So let us compare it with other materials in terms of specific elastic modulus (i.e. stiffness) and specific tensile strength: the word 'specific' in each case implying that the standard cross-section is reckoned in terms of a multiple of unit area and relative density. For carbon, the former index of strength may be 176 gigapascals (see Appendix B) and the latter 1.26, whereas for many alloys of aluminium, steel or even titanium, 25 and 0.2 may respectively be typical. So even when carbon fibre is set in epoxy resin, big gains in strength can be registered.[28] Besides which, carbon fibre well withstands thermal stress and generates an agreeably weak radar signature.

What is more, the next quarter of a century or so may see a grand synthesis of the science of materials: this the result of a confluence of knowledge about quantum theory, wave mechanics and nuclear structures.[29] If so, the implications for vehicles as sensitive to structural economy as are fighter aircraft, in particular, could be most positive. After all, a direct saving of, say, one pound of weight (structural or whatever) can lead to the saving of six or seven pounds through the 'knock-on' effect: less engine power, less fuel needed, further structural economies and so on.

Even now, the indications are that novel materials will regularly comprise upwards of 15 per cent of the all-up weight of new aircraft. Take, for example, the F-16XL experimental extension of the F-16 Fighting Falcon. Its 'cranked-arrow' or modified delta wing is largely composed of carbon composites. Yet as this example rather suggests, the great significance of the said revolution in materials lies in the way they facilitate the application of exotic concepts that, in some cases, were first thought of decades ago. One such prospect in store is reducing fuel consumption by up to 40 per cent by strongly inducing 'laminar flow' in the 'boundary layer', this being the stream of air within 2 or 3 mm of the surfaces of an aeroplane in flight. The basic design of the wing, including how it blends with the fuselage, may be decided with this aim in view. Still, the nub of the matter is to draw air down by means of a dense grid of holes plus perhaps a series of fences, these underlain by a suction conduit. Titanium is eminently suitable for the top surface and a corrugated composite for the underlay.[30]

Among other ways of improving the all-important lift−drag ratio of a monoplane wing are these. More or less vertical wing-tip winglets may serve to break up vortices and so diminish drag by maybe a twelfth. The whole wing may be made 'subcritical': meaning that nowhere will the air flow exceed the speed of sound. Alternatively, one may accept that on, say, an air-superiority fighter the wing will be supercritical. In which case, everything must be done to ease the passage of the sonic front. A remedy developed by Dr Richard Whitcomb of the National Aeronautics and Space Administration (NASA) involves a flattened leading edge and a downward curving trailing edge and can cut supersonic drag by as much as 15 per cent.

Among various futuristic ideas first advanced by Clément Ader over 80 years ago is that of wings which are 'variable camber' or, as it is now often put, 'mission adaptive'.[31] The camber of such a wing can be adjusted in the air in accordance with the type of flying required at a given time — subsonic cruise, low-level supersonic dash, or whatever. Another possibility is the 'slew-wing' reinterpretation of the 'variable-sweep' or 'swing-wing' concept already embodied in not a few warplanes. The point about variable sweep has always been that it allows a wide range of speed performance. There can be lowish speeds for landing, thanks to the wings being fully extended; and high speeds, coupled with diminished response to turbulence, when they are fully swept. In terms of the physical principles involved, however, there is no necessity for port and starboard to move either forwards or aft together. So weight and complexity could be eased by the mounting of just one rigid wing on a central pivot.

These last ten years, however, a higher priority for development has been wings that will probably be rigid and, on each side, swept well forward. The benefits can include more choices in regard to aircraft layout; significantly greater lift/drag ratios; lower stalling speeds and hence slower landings; and maybe a lower wave drag even when supersonic. The concept first found expression in the Junkers 287 of 1944 vintage. But the associated weight penalties long remained excessive. Nor is the aerodynamic debate yet concluded.[32]

In its generation, the F-111 variable-sweep deep-strike aircraft (which first entered service in 1967) was the triumphal expression of a long-recognised concept, a culmination made feasible by materials science. Even so, it has been estimated that nowadays the F-111 specification could be met by a machine weighing less than half as much. This would be thanks to the use, in association with the newest materials, of exotic concepts: notably 'forward sweep' and variable-camber wings together with 'variable cycle' engines, ones able to function either as turbofans or pure jets.[33] Needless to say, too, most of the aerodynamic improvements also assist manoeuvrability. However, better responsiveness is always liable to be associated with instabilities that may be dangerous close to the ground. This may apply with the slew-wing, most forward-sweep designs, and — to take a more familiar example — the delta wing. It applies least with pure rotorcraft. Even so, not dissimilar problems could arise with rotorcraft-cum-monoplane hybrids.

But well within the existing 'state of the art' of some design bureaux is what is colloquially known as 'fly-by-wire'. By this is meant recourse to electrical rather than mechanical linkage between the cockpit and the control surfaces. A computer interprets the pilot's instructions in the light of its continuous record of what the aircraft is doing. Duly, it transposes them into the motions required of the ailerons and so on. This approach has come to be known as 'active control' as contrasted with 'passive' — the mode whereby the pilot manipulates the control surfaces directly and piecemeal.

To dispense with mechanical connections is also to save a modicum of weight. Moreover, such saving may be carried further as and when the fibre-optic channels of communication already experimented with (on, for instance, the Boeing YC-14 and the Skyship 600) are routinely installed. But what is operationally more important is that the 'fly-by-wire' solution minimises the physical constraints imposed on a control matrix. Dependability and responsiveness are thereby enhanced. What is more, wire links are susceptible to magnetic interference whereas fibre-optic ones are not.

So the trend should be towards more carefree flying, especially near the ground, of aircraft not otherwise well adapted in this regard. Indeed, there is every prospect of the eventual emergence of aerodynamic forms that are 'control-configured', with their centres of gravity considerably further aft than would have customarily been seen as stable. Already simulations carried out in the United States suggest that a 'Control-Configured Vehicle' (CCV) flying very low, albeit at a modest 350 mph, might engage with guns and rockets each and every surface target along its axis of movement. It may do this by depressing its nose some eleven degrees while maintaining a constant altitude. A warplane of more orthodox design would have to alternate sudden dives and sharp pull-ups, thereby being able to attack but 40 per cent of the surface overflown.

The inference has been drawn that CCVs could destroy 1.5 times as many ground targets as their present-day equivalents while suffering only a quarter the loss rate. In air-to-air combat, they might destroy twice as many enemy while their own loss rate would be but a sixth as great as before.[34] Findings so bold can be questioned, particularly as regards losses. Even so, there is little doubt that control configuration could yield dividends. Among other things, it may be applied to low-flying interceptors to enable them to shoot down still lower intruders.

In the meantime, terrain-avoidance radar (a kind of navigational aid that has already been around for a quarter of a century) shows promise of considerable improvement within a few years. First and foremost, it will be possible to have integral to it perhaps 250,000 square miles of topographical data which the system can continually compare with its radar scan in order to check position and which it can use to pre-set a course for the further side of approaching high features. One benefit should be that the radar need be activated a smaller percentage of the time, thereby reducing the risk of location betrayal. At the same time, the scan may widen from a typical 10 degrees each side of the flight path to maybe 40. Correspondingly, there may be more bank-angle lateral manoeuvre, as opposed to adjustment along the pitch-axis. The upshot should be that the representative altitude for terrain-following will diminish from rather over 200 feet now to somewhere between there and 50 feet, depending on geography.[35] Even as things stand, terrain-following normally ensures an accuracy of within 30 ft in the horizontal fix of position, regardless of range.

In many other ways, of course, the onward march of electronics is significant for aeronautical science, quite apart — that is to say — from its being a major element in actual air warfare. Thus the launching of some 30 navigational satellites by various United States agencies between 1959 and 1974 was capped in July of the latter year by the placing in orbit of the first of 18 satellites in the Global Positioning System or NAVSTAR: a joint services programme that gives automatic navigational fixes to within 30 ft (vertically as well as horizontally) and does so to ground-based, seaborne and airborne systems alike.

Meanwhile, airborne inertial guidance continues to improve with errors in the determination of acceleration as well as in orientation by gyro being progressively reduced. Gyro drift, in particular, has decreased from 0.015 to 0.001 degrees per hour since the early 1960s; and now the developmental aim is at least to preserve such performance while going for simplicity of operation plus reliability. This will firstly be achieved through the replacement of gimballed systems by ones in which both gyros and accelerometers are fixed to the airframe. No matter that — in the case of a combat aircraft — vehicle rotation rates of around 100 degrees per second may sometimes be imposed, at least briefly. However, the outlook is for ring-lasers to take over the gyroscopic task in aircraft and, even more assuredly, missiles.[36] With them, the Mean Time Between Failure (MTBF) may rise to tens of thousands of hours. Advances all round should *inter alia* ensure that navigational accuracies low down the bracket 1.0 to 0.1 miles of error per hour of flight are soon the norm for a whole variety of aeroplanes, even without monitoring by the pilot or external fixes.[37]

Out of the broad development of 'avionics' (aviation electronics) emerge suggestions that machines which are lavishly endowed in this respect — such as the F-111 and the Tornado — are 'all-weather' performers in a rather exclusive sense. The full avionic mix on the Tornado will include a laser unit, a Forward-Looking Infra-Red (FLIR) unit and a doppler radar set along with a digital computer that, in addition to inertial reckoning, can perform contour-matching, steering and weapon-aiming functions. A second computer is attached to the information display, and can be used not only to generate the basic symbology the display requires (see below) but also bomb-impact lines and gun-shell trajectories.

Still, the notion that certain exotic designs are distinctively all-weather can be criticised from several angles. On the one hand, many other warplanes can be tolerably effective in bad weather against quite a variety of air and ground targets. Yet on the other, not even such remedies as radar-scanning and laser-ranging are proof against every state of weather or light. Laser emission, in particular, is useless either for ranging or for the illumination of pre-selected targets if attempted through cloud, mist or thick haze. In fact, certain planes that do carry a laser set are not customarily regarded as 'night and poor weather' machines.

Furthermore, even a plane well able to adapt to cloud, rain or snow during an actual engagement may, none the less, have difficulty taking off or, even more particularly, landing. Yet when the phrase 'all-weather fighter' came into vogue in the 1950s, airfield performance was seen as the crucial criterion. Nowadays, the representative peacetime limits for the acceptably safe landing of a 'fast jet' are not too slippery a runway, a forward visibility of at least a mile and a cloud base not lower than 250 ft. The last mentioned is mainly determined by the persisting tendency for reception by the Ground Control Approach (GCA) radars in airfield control centres to be blurred by 'ground clutter' below roughly that altitude. So even in wartime, these limits should rarely be extended much further. A crash on the runway is the last thing you want.

Another contribution that electronics can make concerns the actual design of aircraft. Computerised calculus apropos such matters as streamlining and stressing is an obvious example. Less obvious, yet ever more important, is the preliminary evaluation of alternative designs through the use of simulators: the static units that recapitulate on the ground the circumstances that confront a pilot in the sky. Ever since World War I, these have been employed in pilot training in some shape or form. But of late the genre has become a lot more sophisticated, thanks not least to the progress being made with digital computing. So no advanced aircraft ever takes to the air nowadays without 'many hundreds of hours of simulation. Many prototype aircraft have been saved by the exposure of dangerous characteristics in flight simulation . . . Many of the military handling criteria are based on evidence from simulators.'[38] At the same time, the rise of several orders of magnitude in half a century in the number of hours spent in full wind-tunnel and flight-testing has started to level out.[39]

To take a particular example from recent USAF experience:

> The F-15, which was designed this way, has had no cockpit changes after two [sic] years in service. At the same stage of its life, the earlier F-4 Phantom, designed without the simulator, had had dozens, most of them very expensive . . . Heavy use of simulation in design is probably why the F-15 was rare among recent combat aircraft in not having a loss in testing.[40]

What then of the manifold evolution of cockpit design? Always the underlying aim is to ease the strains imposed by such modern obligations as 'round-the-clock' operations, electronic warfare, and flight fast and low. Thus the application gathers pace of the concept of Head-Up Displays: the projection of in-flight information (principally, to date, through the use of cathode-ray tubes) onto an inward-facing surface set just inside the windscreen.

Putatively, too, there is an eventual prospect of the actual aiming of weapons being conducted merely through the movement of the head in response to visual observation and in relation to a collimated reticle. Provisional evaluations of this concept of helmet-mounted sights have indicated benefits, both in the air-to-air mode and in the air-to-ground. Nevertheless, the current systems are limited by an obligation to point the head towards the target, hardly an appropriate provision given that fine sighting is normally accomplished by eye movements. Duly, proposals have been made to integrate the motion of the eyes themselves, perhaps through the use of infra-red sensoring to discern changing reflections off the eyeball!

Nor might even this represent the final synthesis of the human factor and the electronic. According to certain enthusiasts, this line of evolution will be concluded only with a by-passing of both limb movement and sensory perception in the translation of stimuli. They aver there can be 'no denying the progress made in this field during the last few years. The present publication rate of around 1,000 papers per year in the scientific literature reflects this interest.'[41] Obviously, one could question so trite an identification of 'interest' with 'progress'. All the same, electrodes pressed on the human scalp do indicate correlations between stimulus and response that might become — via a fast computer — a means of interaction with the equipment being manned. 'The operational worth of such techniques is very open to speculation at this point in time but they may be used for simple logic decisions such as weapon firing.'[42]

Of more immediate relevance, in any case, is the progress in train with small unpiloted aircraft, thanks mainly to improvements in electronics and in turbine propulsion. Some of these vehicles may serve for such non-lethal purposes as reconnaissance. Witness the advent of helicopter drones with rotors that are but two or three metres in diameter. Still, the manifestation most remarked on as of the present is that in the form of the unmanned monoplane bomber we call the 'cruise missile'.

The essential particulars of the current US versions are well-known. Nevertheless, they have become sufficiently central to the whole debate about air power to merit adumbration here. The USN is bringing the Tomahawk into service; and the USAF, its AGM-86B Air-Launched Cruise Missile (ALCM). The latter normally has a launch weight of roughly 8,000 lb and a range of about 1,500 miles. The former weighs about 14,000 lb and may appear in two modes. The one is as a strategic weapon with a range of rather over 2,000 miles and a nuclear warhead with a low-kiloton yield. The other is as an anti-shipping missile with a range of about 750 miles and a non-nuclear warhead of about 1,200 lb. The two front modules, which contain the guidance system and warhead, are alterable in accordance with the specific mission. It should also

The Impact of Novel Technology 81

prove possible to vary considerably the balance within a cruise missile between warload and fuel.

The strength of all these systems lies in the way they can combine impressive range and lethality with high subsonic speeds and pronounced compactness. Thus the shortest variant of the Tomahawk is twelve feet long. With or without Stealth technology, such dimensions are conducive to the projection of quite tiny radar images and infra-red signatures. What is more, this compactness coupled with a minimal need for launch support renders the said weapons eminently eligible for mobile release, this perhaps in considerable number even from a single platform.

Thus a B-52G is able to carry eight ALCMs internally, in a rotary dispenser. In addition, it will be able to mount twelve on external pylons, one advantage being that an auxiliary fuel tank might then be attached to each ALCM. Nor is the B-52 by any means the only plane that could be fitted out to carry these delivery vehicles. Among other types eligible are such light attack aircraft as the A-7 Corsair; the C-141 Starlifter and C-5 Galaxy transports; the P-3C Orion Long-Range Maritime Patrol machine; the S-3 Viking anti-submarine plane; and, reportedly to the tune of no less than 100 of them, the Boeing 747.

A small quantity of a foreshortened version of the Tomahawk might likewise be released from a B-52. Alternatively, and maybe with the aid of booster rockets, it could be dispatched from sites on land. More regularly, one variant or another of the Tomahawk is available for firing from the weatherdecks of warships or else through the torpedo tubes of submarines.

High standard accuracy is already among the most salient attributes of this genre. True, what was said above about inertial guidance does imply that, if solely reliant on its inertial system to steer across the sea or some other featureless surface, such a weapon might drift a couple of miles off course over the span of a couple of hours or so. However, it is possible regularly to have automatic fixes from a navigational satellite. In addition, the Tomahawk will be equipped with terminal-guidance based on active homing by means of a frequency-agile radar which is aided in its ECCM responses and target determination by an on-board computer.

Such a weapon can also adopt a suitably erratic flight path at a height of 30 feet or 40 feet above the sea. Likewise, it can today proceed, at Mach 0.75, at 50 to 100 feet above average terrain or

250–300 feet above that which is rugged or wooded or built over. As with manned aircraft, such margins are sure to diminish over time.

Evidently, cruise missiles such as those just referred to mightily extend the striking power of the aircraft and of the naval vessel. No less evidently, however, other forms of the genre will be direct substitutes for these platforms in that they will be land-launched. The BGM-109A Tomahawk variant is already a famous case in point, on account of the controversial implications for the arms control dialogue of its current deployment in Europe.

Ultimately, this argument may be extended by the entry into service of much longer-range cruise missiles launched from the surface. Some times it is said that the extra fuel required would make an intercontinental 'cruise' too cumbersome. All the same, the US Defense Advanced Research Projects Agency has been working on a proposed Ground-Launched Cruise Missile (GLCM) with a range in excess of 6,000 nautical miles.[43]

Other Advances in Air Ordnance

On the face of it, air power — above all, as manifest in the manned warplane itself — is a singularly potent means of transmitting, over extended ranges, explosive firepower. The weight of the warload that may be consigned in one delivery is great. Likewise, the individual projectiles to be delivered by this means can be made exceptionally large.

Thus the largest Soviet strategic bomber currently in service can carry three times as such explosive (nuclear or otherwise) as does the largest Soviet ICBM, while the corresponding US ratio is no less than nine to one. Then again, the biggest aerial bomb dropped in World War II weighed ten times as much as the biggest battleship shell. Similarly, the bomb most frequently dropped by the USAF during the second Indo-China War was the Mark 82 General Purpose; and just on a third of the 280 lb weight of this gravity or 'free-fall' weapon is the explosive charge itself. Conversely, the shell most frequently fired by US artillery in that campaign — the M1 High Explosive (HE) 105 mm howitzer cartridge — weighs 40 lb, less than an eighth of which comprises its lethal explosive charge.

Yet historically, such differentials in delivered load have counted for a good deal less than might have been expected. This is because the radii of destruction around explosions grow much more slowly, proportionately speaking, than the release of energy these involve. Not that the actual correlation is ever easy to determine or describe. Among the complications is the burst scatter of solid particles: debris, case fragments and perhaps, too, metallic balls or flechettes. Across a wide spectrum of ordnance weights, particles tend to be lethal to humans over a radius at least as great (and perhaps many times as great) as is blast as such. They can also injure much further out, especially when proximity fuses are used to achieve air bursts.

Nor is a blast wave as such at all easy to assess. Deflections apart, its strength at a given time will be a function of four effects that do not correlate all that simply one with another. These are the time a wave takes to pass over; the associated wind speeds; the profile of pressure change with time; and the peak overpressure registered. For the moment, however, it is enough to concentrate on the last mentioned, this being measured in kilopascals (kPa): a *Système Internationale* unit, 100 of which nearly equal an atmosphere (see Appendix B). Thus around 50 kPa, a human being in the open is liable to suffer acute injury; and an aircraft on the ground may well be rendered inoperable. At four times that strength, a human thus exposed would almost certainly face instant death, while many vehicles or buildings would be crippled or wrecked.

A 25 kg charge of standard high explosive bursting on the ground is therefore virtually certain to cause death from blast alone to unconcealed troops 10 m away, as is one of 1,000 kg at about 40 m. All things considered, however, 50 m and 100 m may be closer to the limits of near-certain lethality. In fact, British tables currently in use on firing ranges give 40 m as the radius for a high chance of death from a 105 mm high-explosive shell, and no less than 620 m for a 'slight' chance of being killed by a fragment from it. Correspondingly, a 1,000 kg bomb is seen as posing a significant threat to human life within 500 metres.

As is suggested by the basic comparisons just drawn, the physical relationship between explosive force, distance from detonation and strength of the blast wave obeys something like a cube-root law. In other words, an increase in explosive yield by, let us say, a factor of 64 will occasion only a four-fold increase in the peak overpressure a given distance away. Alternatively, it will yield that same proportional increase in the distance at which a certain peak of overpressure is experienced.

Clearly, the connotations of this scientific inference have not been positive for the classic mode of aerial bombing. For it confirms that a heavier unit load does not spread destruction as much further around its point of impact as might have been expected. Moreover, scaling laws just as tight are applicable to the behaviour of fragments. A rider to add is that this makes inaccurate aim all the more futile. Yet, historically speaking, bombing has usually been conducted with at least relative inaccuracy.

A fair indication of the degrees of inaccuracy generally accepted in the fairly recent past emerges from US experience in World War II and just after. The hypothetical challenge posited by one analysis was that of a strike, made under good conditions, against a gun revetment 8 m in diameter. The chances of scoring a direct hit on this hard-point target during a high-altitude bombing pass were put at only 1 in 10,000. However, in a zoom down to 4,000 ft they improved to 1 in 600; and in aimed dive-bombing to 1 in 300. Yet even with bombing runs at low level (150–300 ft), the ratio might still be 1 in 100. So it took a near to 'point blank' attack, with unguided rockets, to achieve a probability of 1 in 20; and this only if the revetment protruded considerably.

Over the next 20 years or so, these prospects did not improve a lot. Therefore, it is appropriate to set them against recent experience in respect of field artillery guns: a form of firepower with which strike aircraft are customarily, and perhaps increasingly, in competition. Under good conditions, a medium-calibre howitzer engaging targets at a range of, say, 10,000 m (and firing shells of an orthodox kind) ought to be able to register a Circular Error Probability (CEP) of 50 m. In other words, half the shots would fall within a circle 50 m in radius and centred on the aiming point. This means, for instance, that it would stand something like a 1 in 75 chance of actually hitting the hypothetical revetment considered above.

Granted, under conditions more representative of a battlefield (shell unclean, imperfectly laid, etc.), those odds might worsen to 1 in 150 or thereabouts. But against that caveat must be set the reality that a field gun will not infrequently train on targets less than 10,000 m away with the CEP diminishing near to proportionately. Also, a single artillery piece may well fire between 100 and 300 shells in the course of a full day's combat.

So without going for now into the questions of comparative economic cost or relative flexibility of operation, one can say that the field artillery gun has long looked quite an attractive alternative to the warplane as an instrument of attack against hard-point positions relatively well forward. What has to be allowed, however, is that neither field artillery nor aircraft have been very effective against tank formations, historically speaking. Reviewing tank casualties during and since World War II, a British analyst has concluded as follows: 'The effect of HE delivered by unguided direct fire is little changed. A squadron of tanks caught in a heavy 155 mm concentration will suffer considerable damage, almost all repairable.' The implication is that even an exceptionally heavy shell is unlikely to cripple a tank unless it scores an actual hit or at least a very near (one or two metres) miss. As a very general rule, however, the likelihood of registering that result with tolerable frequency has been drastically diminished by the precise fall of shot across the target area not being observed.

But he goes on also to suggest, in the light of past experience, that barely five per cent of Armoured Fighting Vehicle (AFV) losses will be caused 'by aircraft, and of these a good proportion will be repairable'.[44] On another occasion, he has argued, on the basis of a representative sample of land campaigns these last 40 years, that tank casualties in battle caused by air attack is typically about 10 per cent.[45] Most tanks have been destroyed across 'open sights' (see Chapter 5), either by other tanks or else by specialist anti-tank ordnance.

However, current developments may well offer to tactical monoplanes and also to land-based artillery the pinpoint accuracy needed against tanks or other hard and compact targets. Various of the more important of the new weapons home on their target by responding to reflections off it of lasered light that will have been played onto it by a designator that could be either surface-located or airborne. The gains may be dramatic. The CEP of large bombs dropped from medium altitude may still be as much as 250 m if they are unguided but well below 50 m if steered (via their fins) along laser beams.

A variation on this form of 'semi-active' guidance is one in which the sensor in the bomb's nose is 'electro-optical'. A televisually controlled descent to target takes place. This is made automatic as soon as the crew of the parent aircraft are satisfied that the bomb is exactly locked on. An alternative rendering of this theme is the fitting of an infra-red module which can be used at night or in haze against relatively warm target surfaces, albeit with some loss of precision because longer wavelengths are being employed. Moreover, these three homing modes are, in principle, available for self-propelled missiles as well. As much also applies to radar-homing even though wavelength considerations may limit its standard error to 50 m or so, as against maybe as little as 10 m if electro-optics or lasers are employed. Radar-homing tends to be most suitable for stand-off delivery at sea. There the targets are largish, isolated and discrete.

An important operational example of an airborne Precision-Guided Munition (PGM) is the Maverick two-stage rocket. The first version, extensively utilised by the Israelis in 1973, weighs 200 kg. It includes a 60 kg high-explosive warhead of the 'shaped charge' genre, long employed to penetrate tough structures by a contrived focussing of the blast wave.[46] In 1973, Maverick's electro-optical guidance reportedly ensured a 95 per cent success rate against Egyptian tanks in the mainly open landscape near the Suez Canal. Optical magnification, laser and infra-red guidance packages can alternatively be fitted. Launched from a plane flying subsonically fairly close to the surface, this missile may travel between one and ten kilometres. It can, of course, be released from an altitude of 10,000 m and more.

Slightly smaller than Maverick is Shrike, an anti-radar missile with a 66 kg fragmentation warhead. In service with the USAF and USN since the 1960s, it works on the principle of homing in on the enemy radar emission. A Shrike can be fitted with one or another of more than a dozen sensory units, each attuned to a different range of frequencies. Alarm, a more recent British development along similar lines, should be able to loiter by parachute at 40,000 ft (c. 13,000 m), while scanning for target emitters.[47] At the same time, Raytheon's use of a minute explosive charge to steer a 40 mm shell in full flight reminds us that the evolution of the artillery gun is by no means concluded.[48]

Nor is the veritable revolution now under way in ordnance just a matter of precise guidance. On the contrary, its most distinctive aspect has been a spreading of the impact of the big warloads, a trend that is particularly enhancing the material and psychological impact of air power. One remedy is to make them cluster weapons. Thus France's Thomson group:

> has developed an anti-tank bomblet from the shaped charge of the well-known SNEB 68-mm rocket. The bomblets can either be ejected from multiple cartridge launchers or dispensed from

cluster bombs such as the Matra/Brandt *Belouga*. The anti-tank version of the *Belouga* which weighs about 300 kg, contains 151 bomblets each weighing 1.25 kg and capable of penetrating 200 mm of armour . . . When released at a height of 250 feet at an airspeed of 450 knots, the bomblets cover a ground area measuring 120 metres by 45.[49]

Suffice to note that the thicknesses of armour plate characteristic of Armoured Fighting Vehicles range from 5 mm to 35 mm in armoured cars, armoured personnel carriers, self-propelled guns or light tanks and from 25 mm to 175 mm in Main Battle Tanks (MBT). However, figures cited for the penetrative capacity of weapons almost invariably refer to standard plate, whereas sandwich and spaced varieties may be two or three times as resistant to most ordnance.

But as is well recognised, the principles of clustering and scattering can as well be applied to minelets. Moreover, the tracks and the relatively weak undersurfaces found even in MBTs are peculiarly vulnerable to this threat. Monoplanes, helicopters and artillery guns may all be suitable for the aggressive dissemination of mines or minelets.

No doubt a moral objection could be raised against increased reliance on any form of mine warfare. In fact, however, concern of this sort has thus far been directed elsewhere. One subject of attention has been the bombs or other ordnance systematically designed to exploit (with balls or flechettes or whatever) the flying-particle effect alluded to above. Just as controversial have been those thickened derivatives of petroleum for which napalm is the generic term. A typical bomb laden with Napalm-B, a variety the USAF adopted in the 1960s, throws out an ellipse of incendiarism perhaps 200 m by 30 m. Temperatures quickly surge to about 1,000° C; and the burning continues five or ten minutes. An MBT thus smothered would effectively cease to exist.

Since 1960, the exothermic reactions of certain other hydrocarbons have been exploited for their blast rather than their incendiary effect. Already, the cumulative blast potential of these Fuel Air Explosives (FAE) can be three to five times that of high explosive of equivalent weight; and, in due course, this ratio may approach ten. On the other hand, the comparatively low overpressure peak achieved by at least certain sorts of FAE limits their effectivness against the harder targets.

What then of the multiple threat to a relatively soft and extensive target area? One study stipulated a requirement to damage severely by air attack each and every one of a random spread of 20 buildings in a logistic area 600 metres, across the sort of target favoured by modish interdiction doctrines. Using 1,000 kg free-falling bombs, an aggregate warload of 100,000 kg would be needed. But if 'smart' (i.e. guided) bombs were dropped instead, this mass could be cut by three-fifths. Were there recourse to air-to-surface missiles armed with ordinary HE, only 10,000 kg need be delivered. Were they to be armed with FAE, just 5,000 kg would suffice.[50]

Still, such possibilities do not exclude gun or rocket fire aimed directly at the surface from low altitude by strike fighters. Take, for example, the 100 mm Brandt aircraft rocket, a device averaging 40 kg in weight. One of its six alternative warheads can penetrate up to 600 mm of standard armour and another nearly two metres of standard concrete. With all such weapons, however, tolerable accuracy remains hard to sustain at slant ranges above 1,000 m. Meanwhile, solid shot from the machine-guns still widely carried on light planes and helicopters may penetrate between 5 mm and 30 mm of standard armour, depending on the actual calibre and range. Against MBTs, the threshold of appreciable effect today lies somewhere between the 20 mm aerial cannon-shell and the 30 mm, the latter possessing a good six times as much energy (kinetic plus chemical) as does the former. The only caveat to enter is that the tips of the shells in use might consist not of the usual steel but of depleted uranium, which is two or three times as penetrative, or else of tungsten, which offers almost an equivalent increase in toughness without engendering political anxieties or technical storage problems.[51]

But quite a critical constraint is that a plane travelling at, say, 450 mph and equipped with, for instance, a pair of the British 30 mm Aden cannon could dispatch only about two such shots in the time it would take to traverse the length of an enemy tank; and even were it equipped instead with an American GAU-8/A (the 30 mm cannon, with six barrels revolving 'gatling' style, now installed in the Thunderbolt 2 ground-attack machine) the delivery of rounds in the said interval would be but three or four. Allowing for outright misses, too, one has to question the utility of any aerial cannon-fire against MBTs that are dispersed at all widely and irregularly.

Nor may a solution lie in going for cannon of larger calibre. On the contrary, their inherent unwieldiness could complicate matters acutely. In World War II, the British fitted the 2C variant of the Hurricane fighter with a pair of 40 mm cannon intended for 'tank-busting'; and the Mark 18 version of the Mosquito light bomber with a 57 mm one for maritime and littoral strikes. Yet neither departure was to commend itself much to the RAF tactical staffs concerned. And in the modern context, the gatling concept would be hard if not impossible to blend with these calibres. For air-to-air combat, indeed, even 30 mm is often regarded as uncomfortably large.

Against more lightly protected vehicles, the chances of a 'one-shot kill' would always be much higher. What should also be pointed out, however, is that few land vehicles are as penetrable as are aircraft themselves. In fact, it is by no means unknown for their metal 'skins' to be barely a millimetre thick around the fuselage, and for even the skin that envelops the wing to be no more than about 10 mm. Thus even on the B-1B strategic bomber (designed, after all, for deep low-altitude penetration) some of the surface is only 8 mm thick.[52] True, a skin will be backed by a spaced succession of stringers or stiffeners, each of which will be something like 20 mm thick. Alternatively, the basic thickness may be well above average because (as with the wing of the F-104 Starfighter) the surface in question is carved out of a solid block. Otherwise, armour may be fitted over really crucial areas, this being integrated to a greater or lesser extent with the main airframe structure.

Yet in spite of these caveats and despite, too, the progressive introduction of new structural materials, the point about structural fragility remains apposite. It does so because of the way modern aircraft designers strive to minimise weight. The F-16 is acknowledged to be one design in which this striving has resulted in great vulnerability to projectiles or their fragments.[53] Among countless historical admonitions is that afforded by the Messerschmitt 109 wreck in which a 20 mm shell:

> had penetrated through the rudder, sternpost . . . various members of the fuselage, the wireless set, two thicknesses of armour plate, both sides of the gasoline tank, the back of the pilot's seat, through the pilot's chest removing various ribs and passed out through the dashboard and out the front of the airplane.[54]

However, that macabre effect must have been registered at point-blank range with kinetic energy at a maximum. For though a 20 mm shell does weigh 100 gm, it has an explosive content of only 10 gm. So even after penetration of fuselage or wing, it normally stands little chance of the one-hit destruction of a modern aircraft, except that a lucky hit on a rotor blade might oblige a helicopter to disengage from the battle.[55]

Furthermore, the question of the effectiveness of gunfire is complicated by the brevity and sheer uncertainty of target acquisition, given that both pursuer and target are moving rapidly and this in three dimensions. Thus USAF analyses indicate that, during the Korean War, an F-86 Sabre typically had to fire 1,000 rounds from its battery of 12.7 mm machine guns to destroy a MiG-15 in the air; and that, over North Vietnam, a Phantom had to fire 500 rounds from its 20 mm gatling cannon to shoot down a MiG-21. With air-to-air missiles, of course, the exchange ratio is almost always a lot lower.

Still, no review of changes in air ordnance ought to exclude the classes of weapon that are adapted to low-level release and dedicated specifically to immobilising aircraft on their respective home bases. Among the earliest developments in this field was the 'concrete dibber' (designed by the French firm of Matra) that reportedly was carried by some 40 per cent of the Israeli planes sweeping low over Arab airfields on the fateful morning of 5 June 1967. A basic principle much favoured for such devices is that of the double charge: the initial explosion serving to enter a concrete structure and the main one then exploding inside or below it. Though the structures principally borne in mind are runways, hardened command centres are also prospective targets.

Penetrations of up to ten metres thick have been recorded in mixes of concrete and earth, the sort of depth that might well be deemed impressive for quite a heavy bomb free-falling from high altitude. Then again, the latest such Matra device is said to be able to go through a 40 cm layer of solid concrete and then shatter utterly, through the 'heave' effect, a stretch of runway or whatever, some 15 metres in diameter. Inevitably, too, it will cause individual cracks that run far beyond the inner core of disruption.

An alternative approach is to make runways and other prepared airfield surfaces unusable by means of clustered bomblets and minelets, the former being intended partly to impede take-offs by an assortment of 'pock marks' and debris while the latter deny access. Perhaps the most advanced application of this principle is Britain's JP233 which will shortly enter service. Each one contains 30 cratering submunitions and 215 area-denial mines; and their interwoven distribution on impact has been the subject of elaborate mathematical modelling.[56] A Tornado may carry two such devices. Just one may ruin hundreds of feet of runway for long intervals of time (see Chapter 9).

Lastly, there is the maritime environment. So far as attacks on surface ships are concerned, recent years have seen a convergence between ordnance launched 'stand-off' from the sky and that dispatched through the air from either on or below the surface. Witness the use of the Harpoon in all three modes. With attacks on submarines, however, stand-off delivery is virtually the preserve of the surface and subsurface platforms. Monoplanes or rotorcraft do better to close with their prey. But whether our anti-submarine weapons are intended for delivery from the air or not, we need to be sure they could penetrate the tough double-hulls some Soviet boats now have. In principle, close-in delivery may assist in this regard. In practice, Britain's Sting Ray appears currently to be the West's only air-delivered torpedo able to meet the criterion just stipulated.

References

1. *Flight International*, vol. 125, no. 3899, 28 January 1984, p. 369.
2. Lieutenant-Colonel B. W. P. Adams, 'Electronic Warfare Given Teeth', *British Army Review*, no. 75, December (1983), pp. 61–3.
3. R. J. Schlesinger, *Principles of Electronic Warfare* (Prentice Hall, Englewood Cliffs, 1961), ch. 6.
4. Martin van Creveld, *Military Lessons of the Yom Kippur War: Historical Perspectives*, Washington Papers, vol. 3, no. 24 (Sage Publications, Beverly Hills, 1975), p. 31.
5. 'Friend or Foe?', *Flight International*, vol. 127, no. 3956, 20 April 1985, p. 1.
6. Julien S. Lake, 'Friend or Foe: You'd Better Know', *Defense and Foreign Affairs*, vol. 6, no. 8, August (1978), pp. 39–40.
7. W. W. Momyer, *Air Power in Three Wars* (Department of the Air Force, Washington DC, 1978), p. 149.
8. 'NATO Identification System', *International Defense Review*, vol. 14, no. 2 (1981), pp. 176–7.
9. *Flight International*, vol. 120, no. 3773, 29 August 1981, p. 623.
10. *Aviation Week and Space Technology*, vol. 114, no. 6, 9 February 1981, p. 69.
11. *International Defense Review*, 10 (1983), p. 1419.
12. *Aviation Week and Space Technology*, vol. 120, no. 14, 2 April 1984, p. 63.
13. *Aviation Week and Space Technology*, vol. 120, no. 21, 21 May 1984, p. 182.

14. *NATO's Fifteen Nations*, vol. 26, no. 2, April–May (1981), p. 76.
15. V. D. Sokolovsky, *Soviet Military Strategy* (Military Publishing House, Moscow, 1965), ch. 5.
16. V. Mihailov, *Soviet Military Review*, 8 August (1972), p. 9.
17. *Defense and Foreign Affairs Daily*, vol. 9, no. 164, 25 August 1980, p. 1.
18. *Aviation Week and Space Technology*, vol. 114, no. 4, 26 January 1981, p. 18.
19. *Aviation Week and Space Technology*, vol. 120, no. 9, 17 February 1984, pp. 16–18.
20. *Computer Weekly*, 30 August 1979.
21. R. H. Lange, 'Design and Concepts for Future Cargo Aircraft', *Astronautics and Aeronautics*, vol. 14, no. 6, June (1976), pp. 24–32.
22. John E. Allen and Joan Bruce (eds), *The Future of Aeronautics* (Hutchinson, London, 1970), p. 224.
23. John Allen, 'The Next 75 Years in Aerospace', *Flight International*, vol. 125, no. 3898, 21 January 1984, pp. 166–71.
24. Bill Sweetman, Pierre Condom and Nick Cook, 'Are propellers coming back — and when?', *Interavia*, 6 (1985), pp. 615–20.
25. Peter Middleton, 'Frugal, fast but fractious', *Flight International*, vol. 126, no. 3921, 18 August 1984, pp. 113–35.
26. Graham Warwick, 'Materials are the key', *Flight International*, vol. 123, no. 3845, 15 January 1983, pp. 146–7.
27. W. A. Stauffer and J. H. Wooley, 'Future trends in aircraft structural materials', *Interavia*, 3 (1979), pp. 212–14.
28. W. G. Molyneux, 'The Potential of Advanced Technology for Aircraft Structures', *Aerospace*, vol. 5, no. 7, September 1978, pp. 17–24.
29. 'Einstein's Unfinished Revolution', *The Economist*, 10 March 1979.
30. 'Laminar Flow a Reality', *Flight International*, vol. 121, no. 3810, 15 May 1982, p. 1208.
31. Oliver Stewart, *Aviation: the creative ideas* (Faber and Faber, London, 1965), p. 31.
32. B. R. A. Burns, 'Forward Sweep: the pros and cons', *Interavia*, 1 (1985), pp. 39–41.
33. *Flight International*, vol. 119, no. 3763, 20 June 1981, p. 1951.
34. 'The Next Combat Aircraft Fly Sideways', *The Economist*, 16 December 1978.
35. *Aviation Week and Space Technology*, vol. 117, no. 2, 12 July 1982, pp. 73–5.
36. F. R. Foggie, 'Advances in Inertial Navigation', *Electronics and Power*, vol. 24, no. 8, August (1978), pp. 582–4. See also 'Why Lasers?', *Flight International*, vol. 127, no. 3950, 9 March 1985, p. 27.
37. *Proceedings of the Conference on Aerospace Electronics in the Next Two Decades* (Royal Aeronautical Society, London, 1979), Paper by G. E. Beck, Fig. 7.
38. K. J. Staples, 'Current Problems of Flight Simulators for Research', *Aeronautical Journal*, vol. 82, no. 805, January (1979), pp. 12–32.
39. For a graphical study of the trend with aerial transports, see *Flight International*, vol. 125, no. 3897, 14 January 1984, p. 89.
40. *The Economist*, 1 July 1979.
41. *Proceedings of the Conference of Aerospace Electronics in the Next Two Decades* (Royal Aeronautical Society, London, 1979), Paper by J. W. Lyons, p. 11.
42. Ibid.
43. *Flight International*, vol. 122, no. 3822, 7 August 1982, p. 290.
44. Major D. E. King, 'Battle Damage to Tanks', *Journal of the Royal Electrical Mechanical Engineers*, no. 31, April (1981), pp. 43–8.

45. 'The Survival of Tanks in Battle', *RUSI Journal*, vol. 123, no. 1, March 1978, pp. 26–31.
46. R. F. Bather and N. Griffiths, 'Some Historical Aspects of Shaped Charges', *Army Quarterly and Defence Journal*, vol. 114, no. 2, April 1984, pp. 190–201.
47. *Flight International*, vol. 123, no. 3848, 5 February 1983, p. 803.
48. 'Manoeuvrable Projectiles', *Military Review*, vol. LXV, no. 7, July (1985), p. 76.
49. *International Defense Review*, 8 (1984), p. 794.
50. J. W. Cane, *Orbis*, vol. 22, no. 1, Spring (1978), pp. 217–26.
51. Peter K. Johnson, 'Tungsten versus Depleted Uranium for Armour-Piercing Penetrators', *International Defense Review*, vol. 16, no. 5, May (1983), pp. 643–5.
52. *Aviation Week and Space Technology*, vol. 119, no. 5, 1 August 1983, p. 52.
53. *Aviation Week and Space Technology*, vol. 106, no. 15, 11 April 1977, pp. 19–20.
54. J. Tanner (ed.), *Fighting in the Air* (Arms and Armour Press, London, 1978), p. 273.
55. *International Defense Review*, vol. 10, no. 12, December 1977, p. 865.
56. Mark Hewish, 'The JP 233 Low-Level Airfield Attack Weapons System', *International Defense Review*, vol. 17, no. 4, April (1984), pp. 485–8.

5 THE CHANGING LAND BATTLE

Movement and Fire

The advent of Emerging Technologies (ET) strongly encourages perceptions of a surface-to-air-space continuum rather than the classic separation between land and air and sea. Even historically speaking, this triadic model has been inimical to a proper interaction between air and surface arms. For one thing, it has encouraged the historians of land campaigns to treat 'the air' as another background topic, along with the weather, logistics, etc. Needless to say, too, these several factors are liable then to be neglected in favour of such glamorous themes as mobile armoured warfare or inspired generalship.

Yet to talk instead of a 'continuum' is not entirely adequate. There is a sense in which the control of the surface and especially the land surface, is more fundamental than is control of the skies. Tactical aviation has regularly to justify itself in terms of its impact on the land battle. The converse is far less the case. In the ultimate, it is the soldiers who win ground or hold it or lose it.

All too often, however, discussion of the future land battle is too generalised to inform satisfactorily the air power debate or very much else. In particular, it fails to allow for the geographical singularities of different theatres: meaning, first and foremost, the number of troops on the ground in relation to the width of the front. Yet these singularities considerably determine what scope exists for air power and what forms thereof may best exploit it.

Over and beyond which, however, two broader questions present themselves. The first is whether the overall thrust of the new technological revolution is in favour of defensive warfare, irrespective of the air dimension or of whether the two sides vary in weapons quality. The second is whether, given the new technical opportunities, whatever measure of qualitative ascendancy the West may still enjoy might be turned to greater advantage than before.

Gains in overground mobility that are remarkable by any previous standards are inherent in the said revolution. Yet much as is the case in the skies above, these are a minor theme in comparison with the advances in firepower now being registered. Indeed, among the more singular developments are two which are able to turn overground mobility into a source of vulnerability rather than a means of evasion or exploitation. Thus the discernment of moving targets by means of doppler radar (see Appendix A) continues to get more proficient along the ground as well as in the air. Likewise, the scatterable minelets now entering service could readily be sown across or around moving columns, so as to ensure they cannot further proceed without placing themselves in peril.

Still, one respect in which both fire and movement are making strong gains is surveillance by night. Only 30 years ago, the most advanced aids were clumsy infra-red emitters, devices all too prone to betray their own whereabouts. Some 20 years ago, image intensifiers reliant on ambient light were entering service. More recently, it has been claimed that the aids to nocturnal vision will soon so improve that 'tanks, mechanised infantry and helicopters will be able to manoeuvre and engage the enemy at night as they can in the day'.[1] Granted, such claims do tend to overpitch. Even so, progress in this sphere may well be sufficient to curb local probing and infiltration under cover of darkness, while facilitating night attacks sufficiently heavy to make the defence feel under pressure 'round the clock'. Naturally, the tendencies thus juxtaposed will make army commanders ever more keen to have Close Air Support (CAS) available during the hours of darkness, be it by planes or rotorcraft.

The Liddell Hart Fallacy

What the spread of night surveillance exemplifies, of course, is how much the revolution in military science now gathering pace owes to applied electronics. Yet the current received wisdom, which is that the potency of non-nuclear defence is markedly on the increase, never arose out of any real awareness of electronics as such. Rather its origins lie in a thesis propounded in 1960, primarily with reference to the European theatre, by the late Sir Basil Liddell Hart.[2] It was that, these last two centuries at least, 'the Defence has been gaining a growing material ascendancy over the Offence'.[3] This pre-electronic trend was, he believed, a consequence of a steady fall in the number of troops needed to hold a mile of front in pitched battle. That in its turn was due largely to successive improvements in firepower though also to ones in mobility and communications. Everything was seen as moving towards the same agreeable consummation.

Alas, this interpretation has to be challenged on a variety of grounds, both logical and empirical. Not the least is that more dynamic night warfare has proved to be the only specific trend within the current revolution that Liddell Hart foresaw; and even on this score, he conspicuously failed to remark the adverse effect of 'round the clock' for the weaker side. Indeed, he nowhere acknowledges that the trend towards combat becoming more dynamic in every aspect *ipso facto* redounds to the disadvantage of the defence. In particular, NATO has long expressed the prospects for defensive war in Germany largely in terms of whether its leadership can or should 'raise the nuclear threshold'. By this is meant extending the number of days for which the alliance could delay recourse to theatre nuclear weapons. Other things being at all equal, greater ferocity of encounter can only foreshorten that interval.

Then again, Liddell Hart disregarded the air dimension in a manner all too typical of the land-oriented military historian. Nor did he address another crucial distinction, that between the lighter and the heavier kinds of land weaponry. The former are often carried and fired by individuals; mainly loaded with solid shot; usually fired at a high rate in quite prolonged bursts; and very generally aimed over 'open sights'. The latter are crew-manned; use explosive shells; and fire plunging shot against targets not in view to the gunners.[4] Conversely, engagement over open sights means that the targets are visible to the gunners. It is normally spoken of as 'direct fire' as contrasted with 'indirect' or plunging fire against targets beyond the line of sight.

Since the early years of World War I at least, the improvements in the potency of the lighter weapons have been less momentous than have those in the heavier. Thus the advent of the caterpillar tractor generated new scope for the application of explosive firepower, while negating some of the power of the machine gun. It is similar to how, in air-to-ground warfare, the accent has moved from the aerial machine guns (so effective against the German army's offensive of March 1918, as shown in Chapter 1) to rockets and bombs.

Both elementary reason and a deal of historical evidence firmly indicate that, whereas greater effectiveness on the part of lighter weapons do favour men resisting frontal attacks from within prepared positions, such a trend in the heavier systems works the other way. Nor is there much doubt that this dichotomy will persist. Let us consider, for example, the incremental threat posed by Precision-Guided Munitions (PGMs) to fixed fortifications, minefields excepted. Basically, this derives from accuracy rather than weight being the chief determinant of efficacy against hardpoint targets. More specifically, it is because, between even the most solid of forts and their surrounds, there are bound to be points of contact (embrasures, casements, periscopes and antennae) which will be peculiarly susceptible to the sort of accuracy achieved by laser-designation from the land or even the air. True, this aspect may not have great relevance to West Germany, given the veritable proscription which there obtains against the erection of fixed defences ahead of actual war. But it does relate to several sectors of interest to the West: eastern Turkey, the Golan and Korea.

Besides, even if one could demonstrate that, in whatever may be the average case, the progress of firepower did favour the defence, this could hardly guarantee its doing so in the context of more open warfare. Here then is a consideration which is germane to Germany. Unless heavily reinforced, the NATO brigades and divisions there would respectively have to cover sectors of front which could be abnormally wide, compared with most past experience. Accordingly, they would be constrained to adopt a mobile pattern of defence, with much reliance on counter-attacks or fighting from unprepared positions. Yet such advantages as may be peculiar to defensive warfare *per se* must relate in the main to more static deployment.

Nor can one ignore the increased shock effect in modern land war: particularly on armed forces which, like many Western ones today, are largely composed of short-term conscripts, and of others with little or no experience of actual combat. Nor is more acute shock simply a function of the feasibility of continuous encounter, day and night. Rather it is a function of increased intensity of attrition, above all by shells and bombs and mines. Strike aircraft afford the most dramatic manifestation of this. But they are not that exceptional. Take, for instance, what might be presumed now to be stable technology, that of the tank gun. In 1945, a US battle tank had to fire its main gun a dozen times to stand a 50:50 chance of hitting the side of an adversary tank 2,000 metres down a test range. Thirty years later, a single round could proffer a higher probability than that.[5]

Evidently, too, psychological as well as material vulnerability would be at its most acute as and when offensive action began with a massive surprise attack. What must also be said is that this menace (about which Liddell Hart was dismissive almost to the point of indifference) is becoming greater for other reasons. Among them is the likelihood that, in central Europe and certain other theatres, it will soon become harder to distinguish between a defensive adversary posture and an offensive one. If so, this will be because of a shift away from the blitzkrieg stereotype whereby an offensive is narrowly focussed on a chosen point of breakthrough, a *schwerpunkt* as the Germans say. For concentrated assaults across open ground threaten again to become as costly to execute as they were between the advent of the machine-gun and the invention of the tank. Accordingly, the standard scenario for the land battle is the more likely to depict a struggle across extended frontages. The principle of mass will then be applied not via convergence to effect a breakthrough but through the early commitment of more or less all available forces to a savage contest of attrition the whole way along whatever line of contact geography allows.

The Implications for Air Power

The implications of all this for tactical aviation are considerable. Assume one is persuaded that, thanks not least to psychological factors, airmen are better able to respond to sudden overground onslaughts than are the ground troops themselves. Then one has an argument in their favour that is bound to grow in importance as the menace itself does. At the same time, however, the more geometrical contention that only air power can concentrate swiftly enough around a point of decision (either to support an armoured thrust or else to check it) may become less compelling if there are no longer just one or two *schwerpunkts*. Arguably, however, it will remain compelling enough. After all, threats of breakthroughs and envelopments can still present themselves on a more local and contingent basis. Presumably, no defending formation will wish to fight a dynamic war in several directions if this can be avoided, perhaps through aerial intervention.

Reconnaissance, by manned aircraft or whatever, well beyond the Forward Edge of the Battle Area (FEBA) could eventually become less crucial to the course of an actual battle. This is because there may be no main thrust to identify and because the enemy may be less disposed to keep a good proportion of his ground combat formations grouped to rearward, available for commitment through a *schwerpunkt* as the fray reaches its climax. On the other hand, forward zones may have to be subject to a lot of aerial surveillance (probably oblique and hence from a highish altitude) so as to allow some forewarning of an abrupt surprise attack. Also a pronounced shift to indirect fire and a considerable one towards damage limitation by continual shifts in position could well place a greater premium on surveillance by aerospace platforms (manned or unmanned) over the forward areas once battle has been joined.

Meanwhile, deep interdiction strikes against lines of supply and reinforcement may appear less profitable, when those lines are well spread laterally and when there is less reliance on reinforcement after battle has commenced. Furthermore, the 'second echelon' that (as shown in Chapter 2) is currently so favourite a target assemblage with the enthusiasts for Emerging Technology (ET) might actually present itself in less strength and with a more blurred profile than they anticipate. In other words, the future may afford less scope for 'battlefield interdiction' than this school of thought suggests.

Not that we should be so bemused by the new dynamic as again to fall, as per 1914, for 'the myth of short war'. When the respective sides are pretty evenly matched substantively and in resolve, a more likely pattern may be as follows. A ferocious and short main battle is followed by an extended period of much less intense conflict or even uneasy truce. Then this, in its turn, could be succeeded by one or more repetitions of the sequence. Needless to say, any period of 'stand-off' or 'stand-back' would be considerably due to severe physical wastage and psychological exhaustion; and hence to an urgent need to recuperate, re-equip and reinforce. However, it might also owe much to (a) fear that the military situation would otherwise slide right out of control, in nuclear as well as in other respects, (b) the risk that any large-scale hostilities in central Europe might soon induce an international financial collapse, and (c) a perception by the fascisto-Marxist elements in Kremlin opinion that a 'no-peace; no all-out war' dialectic would help them keep control of Eastern Europe, not to mention the USSR itself.[6]

But a stand-back situation could prove to be one in which most of the roles customarily played by tactical aviation became of greater moment again. What is more, it could favour the manned aircraft reasserting itself as a prime instrument for their fulfilment. This is because the opposing land forces would then be less dense and organised, not least in respect of anti-aircraft work. Clearly, however, this puts an added premium on the adequate survival till then of air squadrons and their base facilities.

Besides which, there are certain more particular trends within the land battle that work against the manned aircraft. One may be that the tank as a genre will no longer be able to preserve adequate immunity through ever-heavier applications of armour. That antidote is (a) largely irrelevant to coping with minelets and (b) unlikely to keep out the more advanced projectiles. Therefore, the inclination must be to settle for fairly light armour. This in its turn will favour a general reduction in the calibre of the main guns borne in battle tanks. For the chief determinant thereof has customarily been the physical strength of adversary tanks. Following that line of argument through, you end up with MBTs much lighter overall. These individually present lesser targets to aircraft. They may also be more dexterous in engaging the latter themselves.

The Triumph of Artillery?

With this diminution of the MBT, individually considered, will go a strengthening of artillery. For one inference must be that field artillery stands to gain at the expense of the tank. But another is that it does so at the expense of the tactical aircraft. This is partly for intrinsic technical reasons. But it is also because of lessened concern with the concentration of firepower on one focal point.

Already, in fact, a reassertion of artillery can be discerned. Among several indications from within the Soviet army is that tank regiments 'now have an artillery battalion, adding integral gunfire support where none previously existed'.[7] More strikingly, Israel's field artillery arm nearly tripled in strength between the eve of the 1973 war and the summer of 1981. Also, the Lance surface-to-surface missile is now being procured from the USA. Also the much longer-range Pershing 2 has been requested, albeit unsuccessully to date. These two national examples offset one another instructively in that, in the Soviet case, expansion is from within a deeply-ingrained artillery tradition, whereas in the Israeli it is from a weak one.

The technical factors just alluded to have thus far essentially been incremental improvements in gun range, rates of fire and so on.[8] Now, however, progress is becoming more a matter of 'quantum jumps' in performance, these perhaps being associated with radical innovations. Not the least among them is the emergence of the surface-to-surface missile (ballistic and cruise) as prospective instruments for the delivery of high-explosive warheads as well as the nuclear or chemical kind.

Moreover, a still more revolutionary departure is in train within the United States. That is to say, the development is in hand of artillery shells which, as they descend to a target area, respectively separate into several capsules, each programmed to home onto a tank or some other kind of 'hard point' target. What this is tantamount to, of course, is a situation in which, so long as its general location is known beforehand, a mobile target can be aimed at individually without either the artillerist himself or anyone observing for him being in a 'line of sight' relationship with his prey. Verily, this could prove to be the gunner's equivalent of the philosopher's stone, especially if combined with the shock effect of the Multiple-Launch Rocket System (MLRS).[9]

Still, there is one respect in which any system that delivers ordnance across any distance ballistically (that is, over a free-throw flight-path) has become a lot more vulnerable. Radar can now determine such a trajectory and so, by backward extrapolation, ascertain the point from which the projectile in question was fired. The only way to prevent this might be to introduce erraticism into the early stage of ballistic flight, probably by means of auxiliary rockets.

One inference to draw is that, come the turn of the century or soon thereafter, the Lanchester Square Law will be most purely or directly applicable to duels between artillery guns. On the one hand, the growing potency of guided missiles will often inhibit moves by tank or air formations against one another. On the other hand, improvements in surveillance and, most dramatically, target registration by 'indirect fire' will make 'counter-battery' exchanges far less circumscribed than heretofore by gun protection and concealment. Accordingly, the side weaker in this field may be overwhelmed more quickly. Herein may lie a strong argument for defending forces continuing to place much reliance on aerodynamic delivery vehicles, ones not so susceptible to backtracking.

A big uncertainty at this stage in the debate is the extent to which the new technologies will lead to a thinning out of the forward troops on the various battlefields.[10] Will the most critical factor become the numbers of men available with the relevant training and aptitude? Will the upshot of everything be that a future battlefield would 'be much emptier than has been the case during the last 50 years'?[11] Such a prospect would alter the specifics of target engagement for both artillery and aircraft. Furthermore, drastically altered notions about the normal density of troops on the battlefield are bound to affect broader judgements about the level and character of the fire support needed.

Geostrategic Perspectives

Such unresolved questions preclude meaningful projections beyond the turn of the century of regional land-cum-air balances. So suffice to look at the early prospect in locations of special concern to us. Other things being equal, the potentialities of air power (offensive or defensive) will be greatest wherever offensive land action stands otherwise to make decisive gains most quickly. Needless to say, the weaker the defending side in relation to the attacker, the more promise or the more menace offensive action will hold. But other factors conducive to it include a low density of troops to linear space throughout the battle zone. Then again, a limited area behind the defended front can also be contributive. Unfortunately, this last applies to several of the areas the West is concerned to protect around the Eurasian periphery.

The NATO front in Germany is, in fact, a classic example of a sector that is not merely uncomfortably extended in relation to troop levels but is also lacking in depth. Thus the mighty river Rhine (terribly hard to retreat across, especially if under air attack) lies a mere 90 miles from the inter-German border, down the inviting Fulda gap. Yet once Warsaw Pact troops had broken through to its eastern bank, they would enjoy a decisive advantage in lateral manoeuvre. Worse still, their governments would be left in a devastatingly strong bargaining position in the event of either a formal ceasefire or else a stand-off.

Alternatively, let us suppose the front had not broken and that the relative erosion of military strength had not been too unfavourable. This would probably be partly because NATO had been trading a certain amount of space for time. Yet it has surprisingly little scope for doing this at all, without leaving some important districts in Communist hands. Towns and cities within 30 miles of the border include (from north to south), Lübeck, Hamburg, Braunschweig, Göttingen, Kassel, Schweinfurt, Coburg and Bayreuth. But one difficulty about the German theatre arises out of the front being wide and closely bounded at both ends by neutral territory. The problem is that planes striking deep therefore find it difficult to outflank while so doing the enemy's forward troops and their dense anti-aircraft umbrella.

In north Norway a localised conflict would be too uneven to continue long at all without nuclear recourse. So the issue essentially becomes one of moving in (mainly through airfields) a multinational force sufficiently large and well-balanced to be a credible manifestation of moral commitment. Tactical air squadrons have a special part to play here because of their speed of arrival and because they bespeak an escalation threat. The extreme exposure of the local Norwegian airfields may matter not too much if the aim of deployment is to bring the territory in question more firmly within the embrace of NATO's collective deterrent.

Around the Persian Gulf, the military prospect is very similar, subject to two qualifications. The one is that, under some circumstances, the outlook for land−air defence would be better. The other is that air reinforcements may be obstructed by overflight bans. Meanwhile, the Golan plateau remains as nodal as ever to the Israel−Syria balance. Yet neither side has even the limited depth Israel held in 1973. Therefore rapid and repeated aerial intervention might prove even more imperative than it did then. Likewise, although land forces are densely distributed across the waist of Korea, they again lack depth. This is mainly in respect of the location just 40 miles from what would be the North Korean start line of Seoul, South Korea's political capital and booming centre of industry. Here, too, aerial underpinning is imperative. So how far would this be facilitated by the qualitative ascendancy of the West?

References

1. FM 100−5 *Operations* (Department of the Army, Washington DC, 1976), ch. 2, para. 23.
2. B. H. Liddell Hart, *Deterrent or Defence* (Stevens and Co., London, 1960), chs 10 and 12.
3. B. H. Liddell Hart, 'The Ratio of Troops to Space', *RUSI Journal*, vol. cv,

no. 618, May (1960), pp. 201–12.
   4. Neville Brown, 'The Changing Face of Non-Nuclear War', *Survival*, vol. XXIV, no. 5, September/October (1982), pp. 211–19.
   5. FM 100–5, ch. 2, para. 2.
   6. See Professor Neville Brown and General Sir Anthony Farrar-Hockley, *Nuclear First Use* (Royal United Services Institute for Defence Studies, London, 1985), ch. 1.
   7. *The Military Balance, 1983–4* (The International Institute for Strategic Studies, London, 1983), p. 12.
   8. 'Shoot, Move and Communicate', *Defense and Foreign Affairs*, vol. XIII, no. 8, August (1985), pp. 20–43.
   9. David E. Ott, 'Where the Thundering Cannons Roar', *Defense and Foreign Affairs*, vol. XI, no. 11, November (1983), pp. 29–31.
   10. Don E. Gordon, 'Electronics and the Individual Soldier', *Military Technology*, no. 6, June (1985), pp. 148–58.
   11. Martin Van Creveld, *Military Lessons of the Yom Kippur War: Historical Perspectives*, Washington Papers, vol. 3, no. 2 (Sage Publications, Beverley Hills, 1975), p. 76.

6 TECHNOLOGY GAPS

The Global Sweep

Not only is numerical superiority easier to exploit in the skies than it usually is on the land, qualitative ascendancy is as well. Nevertheless, the latter may be hard to judge except in the exigencies of combat. Is one looking at the technical ambience in the round or at the specific attributes of certain key weapons? And which attributes should be deemed the most important? And how readily can they be measured and compared?

Still, it is only meaningful to discuss such questions fully with reference to a particular theatre at a given point in time. All that can be attempted in a future-oriented review of military aerospace worldwide is a straight comparison of the more evident attributes of the chief weapons and support systems that particular countries are producing.

Here a dichotomy immediately presents itself. Some advanced kinds of system can be independently developed only by a handful of countries. Salient among them to date have been unmanned satellites, Airborne Warning And Control Systems (AWACS) and Precision-Guided Missiles (PGM). Salient 15 to 25 years hence may be ultra-long-range cruise missiles, long-endurance loiter aircraft, manned orbital platforms and military space shuttles. Soon after that, laser guns just might be on that agenda.

Yet on the other hand, some capacity to manufacture and, indeed, design in the aerospace field has lately become more widespread, thus reversing a move to greater concentration from 1945. Czechoslovakia, Brazil, China, India, Israel, Romania and Yugoslavia are among the developing nations that are building military aircraft to their own specifications, even though the engines and other major components are often from abroad. Their emergence thus is in line with a tendency for the technological base of the developing world as a whole (as measured by such indices as the acquisition of motor vehicles or of computers) to double every decade or so. A particularly popular kind of monoplane in countries looking for manufacturing 'lift-off' has been the light strike trainer (see Chapter 12).

106 Technology Gaps

Two nations, Israel and Japan, are rising conspicuously, not just as aerospace purveyors in their own right but as potential contributors to an aggregated Western effort. The former is doing this partly by dint of her experience of dynamic electronic war, notably in 1973 and 1982. The government-backed firm, Israel Aircraft Industries Ltd, was founded as far back as 1953. But not until the middle 1960s was a big expansion of manpower (from 4,000 to 20,000 in a decade) embarked on, with the emphasis switching from servicing to development and manufacture.

Japan's return in force to this scene has been still more recent. The impetus came from a burgeoning tendency for the civilian market to be a pacemaker in respect of certain aspects of electronics and of materials science.[1] Reliance on clear indicative planning has been the key to the remarkably adaptive industrial strategies Japan has pursued since 1945. This planning has been the fruit of the magisterial influence the Ministry of International Trade and Industry (MITI) has exercised. Lately, MITI has placed far more emphasis than before on aerospace development, its declared national aim being to secure 20 per cent of the world market in civil aerospace by the year 2020. Since early 1983, the United States has been actively seeking bilateral technological exchanges in areas related to military electronics. Naturally, Japan's prowess in computer development is of special interest. Neither such exchanges nor the export of sophisticated arms are precluded by the Japanese constitution.

The Central Comparison

Still, quite the most important dialectic in this whole field is, as always, the West–East one: meaning, above all, the rivalry between the USA and the USSR. At first sight, this interaction seems these days to ensure a large measure of parity. To compare a modern Western aircraft with a Soviet one of like genre is to find a very similar mix of systems and techniques. A brake parachute, an In-Flight Refuelling probe, an Identification Friend or Foe transponder, variable-sweep wings, a control for the adjustment in flight of tyre-pressure, a Magnetic Anomaly Detector, a tail-warning radar, a 'look-down; shoot down' weapons set, a laser-ranger, a gatling cannon, an ejection seat and so on are more or less as likely to be found in the one as in the other.

In less banal terms, too, equivalence often appears to apply. Before being surpassed itself in May 1975 by a Soviet E-266M, the F-15 Eagle had clipped by 36 seconds the world record of 244 seconds for a climb to 30,000 metres set by the original E-266 (a prototype of the MiG-25) in 1967. The E-266 had displaced the Phantom, which had made this altitude in 371 seconds in 1962. Three years before, a Phantom had become the first plane to reach the height at all.

No less remarkable has been the equipoise between the Soviets and the Anglo-French in the exotic, though as yet ill-rewarding, field of supersonic commercial transport. A pre-production model of the Tupolev-144 first flew in September 1971, while the BAC-Aérospatiale Concorde did so that December. A commercial service (Moscow to Alma Ata) tentatively began with the former machine in December 1975. The latter started regular passenger flights (to the Middle East and Latin America) just one month later. In various other respects, from configuration to performance to commercial failure, the parallels have been close.

The Tu-144 and also the Tu-22 bomber are driven by a pair of NK-144 turbofans. In all probability, too, the same applies to the Blackjack bomber (see below). Each NK-144 generates (with afterburning) some 19,610 kg of static thrust, one eighth more than what is achieved by the Olympus 593 used in Concorde. Then again, the D-18T has been in Soviet service since 1978. This turbofan closely corresponds to the Rolls Royce RB211, 524 84, which entered service in 1982. As regards the crucial yardstick of turbine entry temperature, in fact, the former is well ahead, even though Soviet designs used to lag badly in this regard.[2] These two comparisons suggest near convergence in gas turbine development almost across the board. Previously, an American lead of several years had appeared to open up.[3]

One sphere of more basic research in which there is a close correspondence now is materials science. In the application of titanium, indeed, the Soviets have acted rather as pacemakers. Their extensive recourse, in the building of military platforms, to this tough and exceptionally heat-resistant metal emanates from the late 1960s. For then the USSR was adopting an advanced method of extracting it from the vast loads of ilmenite ore she mines for making steel. A far-flung export trade in titanium soon followed, along with its widespread employment in military construction (e.g. the hulls of submarines, and the floors of Antonov-22 heavy transport aircraft). Rather oddly, though, an aerial defection to Japan in 1976 revealed to us that the first operational version of the MiG-25 (which entered service in 1970) is only 19 per cent composed of titanium. Contrast this with a good 25 per cent in the F-14 Tomcat and the F-15 Eagle.

Very frequently, however, Soviet design philosophy is conservative. What the designers often do, in fact, is rely on belated imitation. Most of the present generation of Soviet warplanes 'could have been submitted to meet US Air Force requirements drafted in the late 1960s or early 1970s. Size, performance and equipment correspond with almost unbelievable fidelity.'[4] In several cases, indeed, particular Soviet models are visibly analogous to earlier American ones. The Antonov-124 transport, expected in service in 1987, closely resembles the C-5A Galaxy operational ever since 1969. No doubt so great a lag here relates to a debate in Moscow about strategic priorities. But take air superiority fighters, assets that any large air force must have. The Sukhoi-27, operational since 1984, looks similar to the F-15, in service a full decade earlier.

To an extent, of course, emulation of Western design objectives in regard to such attributes as load-bearing and duration will be at the expense of such traditional Soviet virtues as rough-field operation or the execution of tight turns. Thus the Su-27 is believed to have a wing loading 10 per cent above that of the F-15 though with a thrust-to-weight ratio perhaps 5 per cent less. On both counts, the F-15 is the plane that should turn the more tightly.

Still, the outstanding exception to replication in design philosophy is the USSR's newest strategic bomber, code-named Blackjack by NATO and expected in squadron service in 1986. It is generally thought to have a loaded weight of 360 tons: virtually twice that of its American counterpart, the B-1B. One factor in this disparity is the Blackjack's being obliged to carry considerably larger cruise missiles.[5] That obligation, in its turn, is largely due to a persistent lack of talent apropos miniaturisation. The Soviet Union still trails the West in such relevant needs as small jet engines, compact navigational devices, and maybe small nuclear warheads.

The likelihood is, in fact, that she will continue to lag appreciably in most of those applied sciences which bear upon aeronautics. Thus although she may currently be coming more on terms as far as aircraft design is concerned, she will probably backslide again in the era of Control Configured Vehicles, forward-sweep monoplanes, and — in the further outlook — such utterly exotic forms as twin fuselages and 'ring wings'.[6] This will be akin to the way the first USAF variable-sweep machine, the F-111, entered squadron service in 1967 whereas the first Soviet, the MiG-23 and Su-17, did so in 1971.

Electronics Again

Be that as it may, however, the most crucial sector in the years ahead is likely to be electronics as applied to surveillance, control and target engagement or, of course, the frustration of adversary endeavours in these several directions. Yet here again less facility at miniaturisation can not only be a source of Soviet weakness in itself but may relate to lags of a broader conceptual kind. Witness the initial slowness of the USSR to exploit the diverse possibilities of phased-array radar. Undeniably, however, the great proliferation of Soviet military electronics that began around 1970 did usher in marked improvements in aggregate capabilities. Take, for example, the maximum effective ranges of airborne radars. The Soviet norm for target tracking was adjudged to have risen from 8 km in 1960 to just over 50 km in 1975; and that for search, from 20 km in 1960 to 100 km by 1975.[7]

Where big improvements in performance are urgently needed by the Soviets is in precision ordnance, a factor of growing importance as a determinant of qualitative advantage. But throughout this sector she is plagued by her customary difficulties in regard to miniaturisation. Thus the CIA is currently allowing only CEPs as poor as 250 m to the SS-18 and SS-19, the two ICBMs that were first deployed in 1975.[8] But that seems no better than what the US was managing with long-range missiles 20 years ago. Now the Pershing 2 (the maximum range of which is 1,800 km) reportedly achieves a CEP as tight as 30 m.

In principle, low accuracy will make Soviet theatre weapons less effective against such targets as runways than might otherwise have been expected. Still, the Pentagon believes that submunitions for the SS-21 and SS-23 rockets have now been developed, just as they evidently have for artillery shells in use in Afghanistan.[9] Meanwhile, however, the SS-20, which appears to have a maximum range of 5,000 km, is only credited with a CEP of 400 m even in its Mode 2, the version with three independently targetable nuclear warheads.

Maybe a desire not to reveal too much about the efficacy of its surveillance makes the West a little conservative apropos of its quotable assessments of Soviet accuracy. But at all events, the deficiency thus presumed is akin to the situation *vis-à-vis* air-to-air missiles. The remarkably large Soviet AA-6 (see Chapter 10) has been conceded an effective range (even with radar-homing) of no more than 50 km, whereas its American contemporary — the Phoenix — weighs but half as much yet can intercept to 160 km. Related, too, is the large leeway the USSR has to make up in respect of Remotely Piloted Vehicles (RPVs) and drones, not to mention again cruise missiles.

The Soviet approach to the electronic acoutrement of tactical warplanes is well expressed in the MiG-25. Although earlier versions at least have relied much on vacuum valves as opposed to solid state circuitry, a good electronic performance is ensured by the strength and quality of these two on-board systems: a radar that pulses at no less than 600 kilowatts and a computer that can cope with an impressively large and complex workload. From them is said to derive an ability simultaneously to detect perhaps 20 aerial targets and actually to engage four of them. Such a coverage is comparable with what is proffered by, say, a F-14 Tomcat. What is more, a MiG-25 can probably sustain such a performance quite well within a tough ECM environment, thanks again to the sheer power of the radar and the dexterity of the computer.[10]

In 1970, *The Economist* said the Soviet computer stock was an average of 10 to 15 years behind the West in both concept and quality.[11] In 1978, the same source reckoned seven.[12] The previous year, a specialist study, drawing on both Soviet and Western sources, had put the current lag behind the United States at five to ten years, there having been little reduction over the previous decade except perhaps in software.[13] In 1981, in the first edition of a wide-ranging assessment of the Soviet panoply, the Pentagon suggested that, whereas in 1965, 'Soviet development and production of microelectronics and computers was about ten to twelve years behind US capability', nowadays 'the average relative position or "gap" is three-to-five years with a few outstanding developments following US technology by only two years and some problem areas lagging by as much as seven'.[14] The third edition, published in April 1984, was non-committal on this score. But twelve months later, an *Economist* survey virtually recapitulated that 1981 assessment.[15]

The Extended S-Curve

What all such gauges of time lapse ought to be related to is the sigmoid or extended S-curve deemed characteristic of technological evolution. What this connotes is that, within a given sector, the improvement of overall performance will stay gradual for quite a while but then become ever more rapid. Then it will progressively slow down again, as the said technology matures. Both the acceleration and the retardation are liable to start quite suddenly. Clearly, the former will widen the qualitative gaps between countries, whereas the latter will narrow them to vanishing point.

The model in question has long been familiar, not least to students of aeronautics.[16] Among its more systematic renderings was in an OECD report published in 1967. To be more exact, however, that actually identified five distinct patterns, the one just described mostly applying to 'specific maturing technologies'. The others included the sudden jump, most vividly exemplified by a first nuclear detonation. Then there was a sustained and near-to-linear improvement with an eventual levelling off. Aircraft vibration levels were cited. A fourth pattern was the close to exponential progress manifested, throughout their most dynamic phase, by 'functional capabilities in areas of concentrated research and development'. Air speed records were seen as a case in point. The fifth alternative was an indefinite 'exponential increase with no flattening in the considered time-frame'; and, so it was claimed, this is 'exhibited by a variety of functional capabilities, among them the representative maxima for aircraft speeds'.

However, virtually from the moment this report appeared, this last example was to be invalidated by the levelling off, at close to Mach 2.3, of the 'representative' top speeds of warplanes and, of course, supersonic airliners. Similarly, the account of climb-to-height competition given above does not support notions of sustained exponential growth in speed records. Instead, the extended S-curve model is more universally applicable than the OECD analysis allows.

But as its author rightly points out, progress in a particular direction, aircraft speed or whatever, is usually registered via a diversity of increments (more exotic fuel, innovations in engine design, improved wings and so on).[17] None the less, the cumulative result is likely to approximate to a sigmoid curve. Arguably, indeed, the entire realm of human progress is subject to this effect. Never mind that the ultimate transition to a stationary state cannot be harmonious without 'a world movement potent enough to meet the challenges'. Without this, the prospect may be 'catastrophe for the species and the planet, regardless of the long-run reassurance implicit in the S-curve phenomenon'.[18] Social scientists and population biologists can debate this prognosis. What it does for us now is emphasise how fundamental is the S-curve effect.

So our presumption ought to be that the electronic revolution now so central to human affairs, civil as well as military, is proceeding along this sigmoid path. As of the present time, most aspects are at the stage of rapid acceleration, the prime example being computer performance. Even so, there are some important facets, such as the basic structure of radar and the miniaturisation of the silicon chip, in which a levelling out seems to be under way. Nor may Stealth technology have too much potential left in it after these next few years. Nor, for that matter, may chaff or conventional Identification Friend or Foe have a lot extra going for them, beyond this decade. Nor has cryptography even now.

Still, although room remains for debate about the limits of their application, laser beams can undoubtedly develop a lot further yet. Meanwhile, fibre optics is poised for exploitation, not least in the aerospace context. Above all, the progress towards 'artificial intelligence' (as expressed in the fifth-generation computer) has major implications for military aerospace as well as for society at large. At last gallium arsenide seems to be coming into its own as a semiconductor, after 15 years of exploratory research. Using it, the miniaturisation of 'chips' can be carried quite a bit further. Also gallium arsenide is a medium in which electrons travel several times as fast as they do in silicon oxide. Apparently, too, it favours the replacement of electron flows with ones of photons which work even faster. All in all, there is little doubt that successive overlays of development will keep an electronic revolution going through the year 2010, at the very least. It is not an expectation that bodes well for the USSR and her allies if only because the 'enterprise culture' of the 'silicon valleys' of the West will remain hard to recapitulate in the Don basin or Siberia or even the Hungarian plain.

No Grounds for Euphoria

Not that euphoria is thereby warranted. Ever since 1945, alas, predictions made in the West of the future of the technological balance with the Soviet Union have regularly alternated between optimism and pessimism, yet equally consistently been wrong.[19] Granted, prognosis in the electronics field is that much easier now in that the frontiers of progress are moving quite visibly as well as very fast, the visibility owing something to an increasing tendency for the civil sector to spearhead innovation. Even so, it is all too easy to be thrown by the sheer variability of the quality of Soviet technology. Take, for example, the authoritative judgement quoted above about an average computer lag of a decade or more as late as 1970. Several years before that date the USSR had become the first country to bounce radar signals off first Mercury and then Jupiter. Yet these signals could never have been discerned had not the astronomers in question possessed an exceptional capacity for computerised resolution.

Then there is the equalising effect of licit civil transfer and of industrial espionage. The 1981 Pentagon study just referred to suggested that, but for civil transfer, the Soviet electronic lag might have been ten to twelve years.[20] Obversely, when a proposed sale of Sentrys to the Shah's regime in Persia was causing high controversy in the USA, Boeing insisted that to 'reverse engineer' the electronic systems of these Airborne Warning and Control Systems (AWACS) machines would take even a highly competent adversary some four or five years.[21] Very likely so. But a lot of technical intelligence might be gleaned the while. A disconcerting precedent is Stalin's penetration of the West's nuclear community in the 1940s.[22]

Still, there is no denying that even occasional and localised shortfalls in applied electronics and other avenues of technological advance may seriously incommode Soviet military aerospace. Such, after all, has been the fate of Soviet-backed air arms from 'Korea' through the Lebanon war of 1982. Nor should we forget that, in the summer of 1940, a British lead of about eighteen months in the application of radar to air control was worth about 20 per cent more planes in air-to-air combat. The logic of the extended S-curve must be that, other things being equal, such an advantage in terms of timespan would count for far more in 1990, say, than half a century earlier. Indeed, it would be far more valuable in 1990 than in 1970.

114 Technology Gaps

Yet other things may not be so very equal. The offence–defence balance throughout the whole gamut of Electronic Warfare may be as crucial nowadays as the quality trade-off between opposing sides in a specific situation. An air superiority fighter does need to be more or less as good as its opposite number, certainly in one-to-one combat. On the other hand, a Soviet Surface-to-Air Missile (SAM) engaging an aircraft at low altitude does not have to be as good as its Western counterpart might be. Nor does a runway-busting missile. They simply have to be good enough to do their respective jobs.

Sometimes, too, developmental lag can be heavily offset by keeping larger numbers in service or by working to shorter re-equipment cycles. A systematic review of Soviet naval literature has shown that the principle of mass is applied in 'radioelectronic struggle' (*radio electronnaya bor'ba* or REB) as in other facets of war.[23] As regards re-equipment, the official USAF assessment in 1977 was that its own cycle of replacement of electronic systems then averaged two decades. So great a span clearly affords scope for a less sophisticated adversary to offset his deficiencies by more rapid phasing in and phasing out. Maybe no small part of the confusion currently apparent in discussion of the West–Soviet technology balance arises from a failure to distinguish comparisons of the best on each side from comparisons of the average in service.[24]

References

1. 'Japan looks beyond carbon fibres', *The Economist*, 18 November 1982.
2. Bill Gunston, 'Soviet turbofan revealed', *Flight International*, vol. 125, no. 3897, 14 January 1984, pp. 76–7.
3. R. L. Perry, *Comparisons of Soviet and US Technology*, R-827-PR (Rand Corporation, Santa Monica, 1973), Part IV.
4. Bill Sweetman, 'New Soviet Combat Aircraft. Quality with Quantity?', *International Defense Review*, 1, January (1984), pp. 35–7.
5. *Flight International*, vol. 125, no. 3898, 21 January 1984, p. 138.
6. 'Shapes to Take Aviation into the 21st Century', *Daily Telegraph*, 28 December 1982.
7. *International Defense Review*, 2 (1976), p. 195.
8. Dr Steve Smith, 'Problems of Assessing Missile Accuracy', *RUSI Journal*, vol. 130, no. 4, December (1985), pp. 35–40.
9. *Guardian*, 8 January 1985.
10. See, e.g., Bill Sweetman and Bill Gunston, *Soviet Air Power* (Salamander Books, London, 1978), pp. 70 and 130–2.
11. *The Economist*, 12 December 1970.

12. *The Economist*, 29 July 1978.
13. R. W. Amann *et al.* (eds), *The Technological Level of Soviet Industry* (Yale University Press, New Haven and London, 1977), ch. 8.
14. *Soviet Military Power* (US Government Printing Office, Washington DC, 1982), p. 74.
15. 'Computer bang — or whimper' in 'Inside Comecon. A Survey', *The Economist*, 20 April 1985, pp. 13–15.
16. For an early example, see Sir Roy Fedden, *Britain's Air Survival* (Cassell, London 1957), p. 94.
17. E. Jantsch, *Technological Forecasting in Perspective* (OECD, Paris, 1967), pp. 156–63.
18. Richard Falk, *A Study of Future Worlds* (North-Holland, Amsterdam, 1975), p. 139.
19. See the author's *The Future Global Challenge* (RUSI and Crane Russak, London and New York, 1977), ch. 15.
20. *Soviet Military Power*, p. 75.
21. *Defense and Foreign Affairs*, 10, 1977, p. 13.
22. See also, 'Why Silicon Valley Finds it Hard to Keep a Secret', *Financial Times*, 2 March 1983.
23. Commander Floyd D. Kennedy, 'The Evolution of the Soviet Thought on "Warfare in the Fourth Dimension" ', *US Naval War College Review*, vol. XXXVII, no. 2, Sequence 302, pp. 41–51.
24. *Air Force Magazine*, vo. 60, no. 7, July (1977), p. 31.

PART TWO

PRE-NUCLEAR WAR OVER LAND

7 THE CLASSIC MISSIONS

The Geographical Setting

The crux of the matter with air power will always be the ability to sortie into enemy air space for purposes of attack and surveillance. One solution to the problem of how is still the one standard in World War II. Advance at medium altitude and in some strength, reliant on the principle of mass and on the organic complementality of different types of machines: escorts, bombers and now ones dedicated to Electronic Warfare (EW). Few analysts would deny that these days, against opposition at all dense and sophisticated, the very kernel of this stratagem must be the suppression of surface-to-air weapons. Yet this may require 'explicit timing . . . an immense application of artillery and surface-to-surface missiles, conventional bombs and ECM pods, as well as supporting airborne jammers'.[1]

So as was intimated in Chapter 2, the Europeans within NATO have heavily discounted this option. They believe they could better exploit their familiarity with the regional terrain and weather by staying low. They feel they do themselves lack the exotic diversity needed for an armada in the skies. Moreover, they doubt whether anybody could make it viable in a real European war. Until about 1975, however, many Americans were strongly attracted to the armada approach. It was consonant with USAF historical experience of precision-bombing by day. The performance parameters sought coincided quite agreeably with those for force-projection world-wide. The US military have traditionally been confident about the exercise of intricate control.

Yet as of late the Americans, too, have come to feel it is often better (not to say mandatory) to go in really low. Nor could anybody deny that, against dense and modern defences, the 1944 daylight bombing routine of advancing *en masse* at 10,000–20,000 ft would often be tantamount to mass suicide. Otherwise, a lot may depend on the specific penetration tactics selected. So may it on the square law in both its quantitative and its qualitative aspects. So may it on geography.

On particular occasions, however, the prospects for the air-to-surface battle will be largely shaped by the weather. Consider, for instance, the broad zone of territories between 45 and 70 degrees North. Almost everywhere, some interior parts of the continents excepted, cloud cover the year round averages 50 per cent or more.[2]

Over central Europe, cloud is exceptionally prevalent. From December to February inclusive, cover below 1,000 ft is observed 28 per cent of the time at the average Federal German location; and on just about a half of these occasions, the base is, in fact, 500 ft or less above the surface. Likewise, fog figures in 17 per cent of all ground records, the internationlly agreed definition of fog being a horizontal visibility on the surface of not more than a kilometre.

Naturally, things are better in high summer, June to August inclusive. However, the betterment is not what might have been expected. A base below 1,000 ft is still logged in one in ten observations; and at 500 ft or less in a half of these. Fog occurs seven per cent of the time, even during this season.

Other meteorological factors could be mentioned. One which must be is how a prolonged passage through cloud with a temperature below, albeit not far below, freezing point is liable to induce icing: the point here being that cloud within such an ambience is usually composed not of ice crystals but of decidedly adhesive droplets of supercooled water. Throughout Federal Germany, the mean freezing level is around 10,000 feet in high summer yet close to the surface in winter.[3]

A synoptic assessment of the operational implications is far from easy. But taking off from and (still more critical) landing on runways can be hazardous in high-performance monoplanes when the cloud base is below 300 ft or forward visibility less than 1,000 yards. Likewise, a cloud base much below 500 ft can still inhibit air-to-ground strikes severely, especially those directed against targets of opportunity. Granted, recent radar developments in the millimetric range have been remarkably successfully at mist and haze penetration in a localised context, especially with reference to terminal homing by ordnance. But even millimetric radar has wavelengths three or four orders of magnitude more than those of visible light, which makes it a rather coarse medium for scanning for compact targets. Nor can viewing a target on a radar screen ever be quite the same thing as seeing it naturally. Also, absorption by the atmosphere itself tends to worsen quite markedly as one moves through the higher radar frequencies (see Appendix A). For several reasons, too, the smoke of battle is likely to hamper strike aircraft more than it does surface-to-surface weaponry.

All in all then, one can readily accept that army commanders in somewhere like central Europe must 'expect a one-third degradation in Close Air Support (CAS) missions during the December–February time-frame'.[4] On the other hand, the efficacy of anti-aircraft defences may also be reduced by bad weather. Sometimes, indeed, cloud cover may permit a medium-altitude approach to target that otherwise would be inviable. After all, most radar emissions are severely attenuated by dense and icy cloud, as are those of visible light by cloud or mist of any kind.

Across the Middle East as a whole, cloud cover is only about a third over what it is across Western Europe. Even over Korea it is rather less. Meanwhile, throughout the world the cloud cover out at sea seems remarkably similar to what it is over nearby land, except that less of it hangs really low. Presumably, this overall similarity is because the influence of mountains and of convective heating in the latter ambience is offset by the greater humidity of the former.

Still more fundamental to an understanding of the interaction between aerial vehicles and the surface environment is geometry: the science of Earth-measurement. Here the first relationship to establish is the slant range to earthly horizons of a body at a given altitude above a flat surface. For in principle, this distance sets absolute limits on the range of surveillance or engagement of the aerial body from the surface or vice versa. The only caveat to enter for the moment is that temperature or humidity gradients up through the atmosphere may cause 'ducts' that considerably extend electromagnetic wave transmissions. These especially occur near the surface, in or near the tropics, and over sea areas.

But so far as the pure geometry is concerned, the key axiom for our purposes is that, at the limit set by the Earth's horizon, a line of sight is tangential to the Earth's circumference. In other words, it will be at right angles to a radius from the Earth's centre to where the line of sight touches the circumference. Therefore what we know as the theorem of Pythagoras can apply. So let us say r is the radius of the Earth; h the altitude of the object above the Earth's surface; and s the distance to the horizon. Then

$$s^2 + r^2 = (r + h)^2$$
$$= r^2 + 2rh + h^2$$

Hence
$$s = \sqrt{2rh + h^2}$$

The Classic Missions

Table 7.1: Slant Ranges To and From Horizon

(a) Metric units

Altitude	Slant range
50 m	25.24 km
200 m	50.48 km
1,000 m	112.87 km
10,000 m	357.06 km

(b) Anglo-American units

Altitude	Slant range
100 ft	12.24 miles
500 ft	28.02 miles
3,000 ft	67.05 miles
30,000 ft	212.10 miles

Note: This table is an exercise in geometry. It takes the diameter of the Earth to be 12,739 km or 7,913 miles, distances appropriate enough to the middle latitudes that we are very particularly concerned with. Then the ranges calculated are the maximum ones between some point on the surface and an aeroplane or other vehicle in the sky.

Table 7.1 gives some values for s plotted against h. But let us consider a hypothetical though not unrepresentative case where there is, in the forward zone, one anti-aircraft battery every mile. If an aircraft is transitting at 250 feet, it will in principle be visible to such a battery up to about 20 miles away. That is to say, there will be a great deal of overlap of fields of anti-aircraft fire.

Except at sea, however, complete flatness of surface is rarely found or even approximated to. In West Germany, for example, 30 per cent of an irregular terrain is covered in trees while several per cent more is built up. Then again, the extremely tortuous nature of the mountain topography of so much of the Korean peninsula is well reflected in the way the many short rivers follow most erratic courses and are studded with rapids and waterfalls.

Occasionally, as with the incised valleys of the Bavarian plateau, topographical fluctuations shield troops on the ground better than they do intruding aircraft. On the whole, however, the effect works the other way. This is partly because, in any case, the weapon and surveillance capabilities in an aircraft rarely extend their influence as far as the geometrical horizon. It is also because anti-aircraft facilities mounted on hill tops to obtain maximum coverage themselves lie exposed. Think of the ease with which radar stations in such locations have been neutralised on various occasions.

What may sometimes be possible, too, is the infiltration of

enemy air space by single aircraft relying on very low and slow flight to avoid established defences altogether. One may cite the 'flight' (a small unit equipped with several aircraft) of Lysander 'army co-operation' monoplanes formed by the RAF in October 1940 to fly men and materials in and out of occupied France, often by 'hedge-hopping' at speeds below 100 knots.[5] Only twelve Lysanders and six pilots were lost in four years, a record which bespeaks an attrition rate well below what other campaigns of intrusion involved. Today, 250 Antonov-2 light transports (carrying up to 12 men apiece) are thought to pose a North Korean threat to South Korea akin to what several old biplanes did during the Korean war. Even tightly defended Israel has been adjudged in some danger thus, especially if its defences were trying to cope at the same time with a high profile attack.[6]

Surface versus Air

However, this much has to be conceded apropos of infiltration as a tactic. It is that, helicopters sometimes excepted, the sort of aircraft adapted to it cannot carry much in the way of ordnance or, indeed, airborne troops. True, they may have a nuisance value disproportionate to what they do carry. Even so, the full shock effect of air power can be registered only by the commitment *en masse* of large and fast strike machines. Naturally, this effect is bound to be most relevant to Close Air Support (CAS): the lending of direct assistance to troops locked in combat.

What the possibilities are for CAS depends on military geography and other contingent circumstances at least as much as for the exercise of any other aspect of modern air power. Already it has been acknowledged that the presentation of surface targets is very much a function of the weather, topography and overall density of troops. It has also been noted that, other things being equal, a low density of troops all round makes extra demands on air power not least because it strongly favours offensive movement on land. Another way to put this is to say that low density is liable to lead to the front line or FEBA becoming highly fluid and perhaps fragmented.

Still, the resulting problems of target determination may be coped with by the employment of Forward Air Controllers. Arguably, too, fluidity as such reduces the ability of ground troops to shield themselves against air attack. Take, for example, a comment made in 1978 by the then Director of Defence Studies, Royal Air Force. With an overground offensive chiefly in mind, he noted that 'Radar-laid weapons require fields of vision, and no rapidly advancing force would be able to pick and choose at will optimum sites for SAM batteries and associated guidance systems'.[7] Nor is there any denying that this constraint has sometimes affected SAMs and anti-aircraft guns during an advance (or, for that matter, a tactical withdrawal) across country which is close or rugged. On the other hand, certain track-mobile crew-served systems, mainly intended for service well forward, may deploy off the road into operational readiness in roughly half a minute. Moreover, some such weapons can search for and maybe even engage targets while still on the move. A case in point is the Gepard Flakpanzer: a 35 mm twin-barrelled system of Swiss origin that has entered service with several NATO armies since 1977.

In many respects, of course, the potency of surface-to-air defence is growing. Thus target engagement time (from initial alert to commencement of fire) will these days be around six seconds for crew-served weapons properly deployed well forward. Only a few years ago, nine seconds was a more typical interval. The difference may often be critical.

Already, too, proximity fuses can be fitted in shells with calibres down to 35 mm with 30 mm soon being possible.[8] Likewise, these devices are incorporated in many Surface-to-Air Missiles (SAMs). When active, they can be triggered (by the miniature radar integral to each) whenever a target surface presents itself within what has been adjudged a lethal distance — maybe 20 to 40 feet when the intended target is an aircraft. However, they cannot be activated if the ammunition into which they are fitted is to be fired low across the land or sea. This is because they will be travelling too fast themselves to discriminate at all readily between a moving aerial target and either static or near-static surface features.

Meanwhile, a pulse-doppler radar, well able to track aerial intruders against a background of surface clutter (see Appendix A), is regularly integral to modern crew-served guns and SAMs. Likewise, an IFF interrogator and a laser tracker have become the norm. Indeed, these last-mentioned devices may even be fitted in shoulder-fired missile launchers.

Simulations and field-tests have consistently indicated success rates for individual crew-served systems of at least 60 per cent against an enemy aircraft passing within their envelope of fire. In such tests, the transit is assumed to be at a representative altitude and, most likely, a high subsonic speed.

Even so, idealised assessments of this sort are not easy to translate into operational terms. Human frailty comes into play. Incoming missiles or drones or Remotely Piloted Vehicles (RPVs) may have to be engaged as well as manned aircraft. A target may circumvent or underfly the successive envelopes of fire. Nor is forceful defence suppression by aircraft or field artillery ever allowed for in the above percentages. Nor are Electronic Counter-Measures (ECM).

Let us assume, however, that an intruding aircraft is transitting very low in order to minimise the time it spends within each envelope of fire or, most alarmingly, interwoven ones. The effectiveness of ECM against surface defences is bound to be circumscribed by (a) the large radar image an aircraft or even a cruise missile will project at short slant ranges; (b) the ready use of laser ranging across such distances; (c) the minimal time available for interaction; and (d) the improbability that any chaff an intruder may eject low down will stay in the atmosphere for more than several minutes.

Besides, even when defence suppression by ECM is being effective, the defence has a remedy. It is to revert considerably to optical ranging and tracking, at least for close-in encounters in clear weather and daylight. Optical sights which magnify several times are often available. Electronic calculators may provide, from angular velocity alone, 'deflection aims' accurate enough against even a high-speed target. Also, laser-ranging can serve as an adjunct to optical sighting.

A French comparison between optical direction and reliance on radar yielded the following result. Suppose a turret fitted with one 20 mm cannon engages an aircraft as follows. The latter is travelling at an altitude of 100 metres and a speed of 280 metres a second (*c.* Mach 0.8); and it comes to just within 400 metres from the turret. Then with radar, there is a 60 per cent probability of effecting a 'kill', presuming the absence of ECM. Without radar, on the other hand, the percentage drops to somewhere between 23 and 6 per cent, depending on the weather and light. Moreover, the radar mode has no need of forward observers whereas the optical would require an arc of them some 4 km out.[9] What must not be forgotten, however, is that even 6 per cent can be made almost 12 per cent if a return transit is reckoned in. On occasion, too, surface-to-air ordnance can effect some attrition and dissuasion even if fired blind in the general direction of a threat.

So let us now look more closely at guns as such. Throughout the West, the early 1970s saw a reversal, in regard to surface-to-air interception lower down, of a previous trend towards SAMs. Thus between 1970 and 1975 the number of crew-served SAM launchers in NATO Europe rose from 1,100 only to 1,250, whereas that for crew-served anti-aircraft guns rose from 1,100 to 1,900.[10] Very probably, too, such shifts of emphasis will go further a while. For one thing, more compact proximity fuses will be of particular benefit to shells. Meanwhile, slew rates of 45 degrees a second in elevation and 80° in azimuth have become typical for the barrels of such weapons. In other words, the agility of the barrel before firing now goes a long way to match the agility of the missile after.

At the lower end of this scale is the machine-gun, the most familiar calibres in the West being 7.62 mm and 12.7 mm. In many armies, weapons of this sort are either poised on bipods or else set in mutiple mounts, as and when anti-aircraft fire is intended. Take, for instance, the Warsaw Pact's ZPU-4: a quadruple mount of 14.5 mm guns set on a four-wheel trailer. With a total firing rate of 40 rounds a second, this is fairly capable against helicopters and marginally so against slow monoplanes. Targets are acquired optically at ranges of up to 1,200 metres.

One weapon of 20 mm calibre (just large enough for the dispatch of explosive shells) is the Vulcan: a US system that works on the 'gatling' principle of multiple rotating-barrels reliant on a single feed. Originally, this weapon was made just for use on-board aircraft. But it has been installed on a variety of surface platforms: a towed trailer, a tracked self-propelled chassis, a gun emplacement or the deck of a ship. With engagement from the surface, the present target may be less fleeting than in air-to-air encounters. Therefore, a surface Vulcan's overall rate of fire can be well below the 100 rounds a second registered in the airborne version, especially since a muzzle adaptor allows selection of the degree of dispersion of shot. The barrels are orientated optically, though a built-in radar and gyro gunsight compute the deflection of aim or the 'lead', as fighting men prefer to call it.

Rather higher up the spectra of aggregate weight and gun calibre is the above-mentioned Gepard Flakpanzer. Target acquisition (down to the 'open fire' signal and the selection of burst duration) is fully automatic within each firing unit's one vehicle. The proximity-fused shells take less than one second to traverse 1,000 m and less than four to traverse 3,000 m. An intruder flying at 'characteristic' speeds and altitudes is deemed to be exposed to a 35 per cent risk of a hit from a single round if its minimum slant range from a Gepard is 500 m. Correspondingly, it is said that a second or two of well-aimed fire (at nine rounds per second per barrel) should suffice to destroy an aircraft.[11]

Still, ever since the abandonment by the US Army of its big Skysweeper gun some 30 years ago, it has been universally accepted that Surface-to-Air Missiles (SAM) are far superior to guns for engaging targets at extended slant ranges and especially near or in the stratosphere. A good indication of continuing SAM progress at this end of the spectrum is afforded by the Patriot, a US system now entering service for rearward defence in theatres like Germany. A firing platoon has three special vehicles which respectively house a radar set, engagement control and an electrical generator; and usually eight road-mobile launch stations, each with four missiles in storage-cum-launching cannisters. Of these, the only manned station is 'engagement control'.

The phased-array radar performs surveillance, tracking and guidance functions that, in the Hawk and Nike Hercules systems that Patriot is replacing in Germany and elsewhere, require several separate radars. Though a Patriot missile is but 16 ft long, it is credited with a maximum slant range of 100 miles. What is more, the agility and precision of this ordnance is such that recourse to a nuclear warhead would commend itself operationally less often than with Nike Hercules. The minimum range and altitude for target interception can be two miles and 200 feet respectively. The target can be travelling at anything between Mach 0.3 and 2.5. The requisite engagement time is roughly 10 seconds.

There are just two conspicuous drawbacks to the Patriot. The one is that each launch station takes an hour completely to reload.[12] The other is a certain awkwardness when switching from movement in convoy to readiness to engage. For this a platoon requires a span of several minutes, as well as a couple of acres of tolerably flat and open space in which to deploy. Perforce it would take a while longer still to redeploy, even across a short distance; and such delay might sometimes render it acutely vulnerable to suppressive fire. Still, Patriot is among the SAM systems now being accorded some capability against such Soviet surface-to-surface missiles as the

SS-21 and SS-23 (see Chapter 13).

Meanwhile, one aspect of the improving responsiveness of SAMs designed to deploy well forward is the ability of some to dispatch several missiles in several seconds. In fact, France's Shahine system, which is mounted on an AMX-30 tank chassis, fires four at once and immediately directs each to a different target. What is uncertain, however, is how often such missiles will prove fast enough to catch low-flying prey. With the sort of systems here being considered, the initial acceleration is close to 30 g. A missile from a Crotale launcher (from which the Shahine is derived) attains its maximum speed of Mach 2.3 in two seconds; and then takes 16 seconds to travel five miles. In terms of instant reaction, however, that performance is weak compared with that of an anti-aircraft shell which leaves the gun muzzle at between Mach 3.5 and 4.5,[13] having been accelerated along the barrel at up to 25,000 g.

Then there are the lighter SAM devices used by infanteers and the like for their own self-protection. A good indicator of progress here is the Swedish RBS-70. This weapon can be fired from a ground-tripod or a light vehicle or a naval vessel. It uses an optical sight (together with laser-ranger and IFF interrogator) to respond to tracking data fed through from a pulse-doppler radar. One man may carry either the tripod or the sight or else one container plus missile. The radar set, called the Giraffe, is mounted in a small lorry. Its retractable mast can raise the antenna a further 12 m. An RBS-70 can be set up for action, with a one- or two-man crew, in half a minute. The engagement time is 4.5 seconds while a missile's flight to 3 km is 8.5 seconds. Anybody can be trained in the system's use in 20 hours.

Also shoulder-launched rockets are widely available. However, their design is always circumscribed by weight limits and by a need to shield the operator from blast and fumes. During the 1973 war in the Middle East, for example, a single hit by a SAM-7 warhead was rarely enough to down an Israeli warplane, even if registered on its engine exhaust. Moreover, the danger posed to the operator by back blast can be avoided only by the use of an auxiliary charge to throw each rocket clear before main ignition. Thanks partly to the split-second delay this involves, many commentators discount such weapons ever intercepting machines travelling faster than Mach 0.75 or at altitudes above 3,000 feet.

Clearly, the sheer diversity and, in some cases, very low cost of the defensive systems potentially available reduces the scope for defence suppression, whether by outright destruction or by ECM. However, an air strike force does have other remedies. One involves missiles being steered along 'lines of sight' onto surface objectives from aircraft 'standing off' a considerable distance. But notwithstanding the Israeli success with Maverick in 1973 (see Chapter 4), one has to say that this stratagem may avail little if the plane in question has to deliver its warload from within enemy air space. Most suitable as a rule is the maritime environment, the targets then being either surface ships or installations near the coast. A big attraction of stand-off is that aircraft can be designed with considerations other than the low-altitude approach to target set to the fore. A drawback is dependence on clear weather.

Meanwhile, there is an alternative mode of stand-off delivery which may often be feasible over hostile land, not least in poorish weather. It is the projection of ordnance onto the target by a plane coming in fast and low. Take, for instance, the Pegase, a winged dispenser of munitions under development by Brant and Dornier for high speed (probably high subsonic) delivery from altitudes down to 60 m. The maximum launch weight is 1,300 kilogrammes and the corresponding warload up to 760 kg. Powered versions can go 15 kilometres with a low-level launch while unpowered ones may glide up to 8 km. Five of any variant of Pegase can be carried on a Tornado.[14]

As regards the choice of flight profile for an intruding aircraft, it has been conceded above that operation at medium altitude is generally infeasible. Never mind its *a priori* attractions in terms of (a) the scope for mass co-ordination; (b) reduced fuel consumption; (c) easier surveillance; and (d) avoiding the physical and nervous strain of low-altitude flight. In fact, flight anywhere above several hundred feet is infeasible unless either the spread of cloud is protective or else the intruding air arm enjoys great qualitative or electronic superiority.

Otherwise, the critical question becomes whether aircraft should fly quite low (*c.* 250 ft) and very fast or instead go extremely low, with a consequent sacrifice of speed. Two British specialists have expressed the view that 100 ft and 450 knots 'is preferable to significantly greater speeds'.[15] As a rule, however, that may be pushing the case for contour-hugging too hard. Some 15 years ago, the Royal Air Force concluded that, in low-level sorties over Europe, a threshold for survival usually presented itself around Mach 0.85. When speeds dropped below that value, attrition did tend to rise sharply. Yet increases towards and through the sound barrier were felt to proffer little in the way of added protection, unless the terrain below was abnormally flat and open.

Across the years, the RAF has emphasised the virtues of low altitude perhaps more than any other independent air arm. None the less, during the great army co-operation debate of the early 1960s, the then Director of Army Aviation in the Pentagon took things a stage further. He earnestly put the case for flying sufficiently slowly to be able to 'hug' very closely the contours below, a tactic that effectively merges with that of aerial infiltration. He cited, by way of hypothetical illustration, a plane that has been traversing a flattish landscape at some 300 ft but which is then obliged to negotiate a knoll rising to a height of 400 ft or 500 ft above its surroundings. At a speed of, say, 260 knots such a machine should be able to 'contour hug' its way over this obstacle. But at 600 knots (about Mach 0.85), it must lurch upwards so sharply as to spend a most uncomfortable 10 seconds at 1,000 ft or above.

The next model he depicted was an 'incised-meander' river valley. This is a singular though not rare land form (caused by geological rejuvenation) in which a river has retained pronounced meander curves yet dug its valley deep. Aircraft travelling either at 260 or at 400 knots are seen as gaining considerable immunity by following closely the course of the river. At 600 knots, a plane could pursue a path which still meandered with much the same amplitude as the river's but which awkwardly evinced a lag of about half a cycle in relation to it. Therefore, it would be obliged to stay above the valley's sloping sides. At 800 knots, there is little choice but to follow a straight line well above the valley axis, exploiting only the virtues of speed and directness. Here, too, 600 knots comes across as the worst alternative of all.

A third scenario he defines in terms of the time an aircraft remains visible to an observer on the ground. A representative case is held to be one in which this person stands half-way between two objects 50 ft high and 4,000 ft apart. A plane traversing at 200 ft and 200 knots is visible for about 48 seconds; and, of course, the intervals for other heights and speeds are in linear proportion, directly so for altitude and inversely for horizontal speed.[16]

Inevitably, however, all these hypotheses are open to revaluation. Intimations that 600 knots may often be a bad choice could well owe something to the 1957 constraint, then still more or less operative, on the US Army acquiring monoplanes with an empty weight of 5,000 lb or above. In other words, a lack of powerful and fast machines may have disposed the army staff more towards aerial infiltration. Since when, the introduction of such technologies as terrain-following radar has been reducing the opposition between low height and high speed.

Furthermore, the field of view notionally available to that ground observer in the third of the Brigadier General's examples does seem too extensive to be typical of the most relevant environments, large airfields perhaps excepted. Surely in Germany, say, one would rarely be able to see 2,000 ft horizontally from a site suitable in general for anti-aircraft emplacement. As a rule, that field of view would be constricted either by weather or darkness or by a morainic bluff, a copse, a building or whatever.

Hardly less crucial, however, is the aviator's need to register potential targets on the ground and to do so in time to hit them. What particularly have to be borne in mind are the 'targets of opportunity': those the exact positions or even the very existence of which have not been known in advance. The special place these always assume in air power discussions stems from their epitomising claims about the flexibility inherent in the pilot in the cockpit.

Reportedly, the indications from other French studies have been that the maximum range in kilometres at which a pilot might expect to sight, on a clear day, a tactical target set amidst representative European terrain is one-tenth of his own height above the ground, measured in metres. For instance, he ought to be able to pick out when at 100 m a fair-sized objective (e.g. an ordnance dump) 10 km distant. None the less, one authoritative source has insisted that, all things considered, defence staffs would be unwise to assume visual detection ranges in excess of four or five kilometres.[17] No matter that, even from 100 m above flat ground, the geodetic horizon is about 35 km distant.

Nor may this advice be ill-founded. Buffeting by turbulence may blur a pilot's perception, perhaps badly. Obscuration by topography or whatever will often be a most critical constraint. Then again, poor visibility will often supervene by day as well as night, not least because of the lingering palls of battle smoke. After all, many fires can be caused in the course of highly dynamic conflict. Furthermore, anything up to a quarter of all the artillery shells fired may specifically be smoke generators. Nor are the more compact targets, such as small groups of tanks, ever easy to pick out from several kilometres away, not even by a stationary observer and against an open landscape. Nor has camouflage been brought into this reckoning.

Yet a pilot thus encountering a surface target not too far away, though maybe well to either port or starboard, could require a good ten seconds to set up a direct attack. The said interval will be the sum of the following phases: the one or two seconds the pilot himself takes to react; two or three seconds for the controls to respond; several seconds to turn, say, a quarter circle; and perhaps another several to straighten into an attack run. However, a plane travelling at Mach 0.8, say, would cover a kilometre every three seconds or so. What is more, if it made a 6 g turn while sustaining that speed it would thereby describe an arc on a circle some 2.5 kilometres in diameter. In short, time may often be desperately short for the reorientation required.

Also during this manoeuvre and the exit therefrom, a plane may expose itself dangerously to surface-to-air weapons. These may be clustered around the target and have reaction times of perhaps 5–8 seconds. What is more, this risk may be increased by a need to rise above vertical obstacles and by a disposition on the part of many pilots always to climb when banking, this through fear of losing vital vertical lift. Thorough training may curb the more extravagant expressions of this tendency but is unlikely to eliminate it.

All of this seems to commend the approach of having CAS aircraft that are relatively well adapted to the absorption of battle damage but also able largely to avoid it. This they may do by slow though irregular manoeuvre coupled with defence suppression or, better still, by engaging targets largely from the friendly side of the FEBA. Such a philosophy lay behind Fairchild's A-10 Thunderbolt 2: the specialised ground-attack (meaning, above all, anti-tank) plane that entered service with the USAF in 1976. Though it can carry up to 9,500 lb of ordnance (including the Maverick), its mainstay is a 30 mm gatling cannon. Fairchild actually claim that in a single sortie this machine may immobilise up to 16 T-62 Soviet battle tanks by engaging them at ranges of up to 2,000 yards!

Reduced immunity from shot and shell is sought in many ways. There are two engines, positioned in parallel high astern. Flight controls are duplicated as much as possible. Well over a ton of armour has been installed, just over a third of it to project the fuel system though nearly a half to form a titanium tub (13–38 mm thick) around the pilot. Duly, the A-10 should normally be able to withstand a number of hits by standard shells from 23 mm cannon, if not by ones of larger calibre or tipped with depleted uranium.

Shortly after the first squadron entered service, a full-scale army-air field test was carried out at the USAF range at Nellis. In the course of 112 offensive sorties, just five A-10s were adjudged to have been lost, though several more were tracked by ZSU-23-4. This loss rate of just under 4.5 per cent can, of course, be seen as acceptably low in relation to the amount of damage so many A-10 sorties might be expected to inflict.[18]

However, the actual infliction either way would depend exceptionally heavily on the military geography and other circumstances. Correspondingly, one should not be too impressed by the apparent survivability of the A-10 as a brand new system operating, in the instance cited, against old-established ones and doing so across a FEBA far better defined than ever the pilots of warplanes might find it to be in 1995, say.

Nagging doubts on this very score can surely be discerned in the remarkable investment put into trying to preserve the pilot. Never mind that (to judge from analyses of recent wars) less than a fifth of aircraft losses in combat actually originate with the pilot being hit.[19] Never mind either that, in a high proportion of the remainder, the loss of the machine is bound to involve loss of pilot and, indeed, any other crew members. Official anxiety is manifest, too, in the special effort the Pentagon made to limit the life-cycle cost of the large and complex machine in question. Part of the price they willingly paid was a very limited capability at night or in bad weather, as witness the absence of the second crew member always called for by Fairchild. Yet soon the implications of this had to be urgently reconsidered (see Chapter 3).[20]

Meanwhile, we must remember that surface-to-air weapons capabilities could well improve markedly over the next 10 to 15 years, especially in respect of a critical foreshortening of engagement times. Moreover, such progress is likely to be evident on the SAM side of the continually shifting though always complementary balance between missiles and guns. Thus, early in 1980, the concept of the Self-Initiating Anti-Aircraft Missile (SIAM) leapt into prominence through the use of such a weapon to destroy a QH-50 helicopter drone at a slant range of two miles. Clearly, the term 'Self-Initiating' is something of a misnomer. Nevertheless, a SIAM does proceed independently of all surface-based equipment from the moment of its launch; and so may be able to sustain pursuit some way beyond the original 'line of sight'. With this missile autonomy comes as well systems simplicity and cheapness and, above all, a reduced susceptibility to ECM. Installation in submarines is among the options being considered with the arrival in service of SIAMs, in 1995 or thereabouts.

An advance more elementary in nature, yet still with momentous implications, is the evolution of SAMs agile enough to be fired from launchers set in a vertical plane, as opposed to ones which have to be slewed towards the direction the target is in. The former arrangement gives close to all-round coverage; and usually lends itself as well to the alternative discharge of missiles other than SAMs. Here again are found the virtues of simplicity and compactness, as witness the hopes entertained for deployment at sea. At least twice as many launchers can be installed in a given place. Both they and their control equipment can be dispersed in modules around, say, a large warship or merchantman. Alternatively, they can be fired from small surface craft. Indeed, a missile as large as the Seawolf, which has a slant range of 20 miles, might thus be borne by a vessel of but 500 tons. Presumably, too, this concept can, in due course, be combined with SIAM.

Lastly, there is the putative possibility of developing for use in aerial combat, lasers that are lethal in themselves. Already, laser range-finding is being conducted at medium altitude across distances in excess of 15 miles. In 1981, a US target drone was engaged for the first time by a presumptively lethal beam. Except that they can be shielded by special goggles, optical retinae are morbidly sensitive to laser flashes, even when these are brief and of low intensity. Essentially, the only exception concerns that part of the lasering spectrum with a wavelength around 1,500Å, a narrow zone of strong immunity afforded by absorption by the aqueous humour in the eyeball. In fact, this lies within the ultra-violet part of the spectrum.[21]

Yet paradoxical though it may seem at first sight, one has to be even more sceptical about this technology achieving an adequate attrition rate against tactical warplanes than against intercontinental missiles (see Chapter 11). Admittedly, what counts as adequacy could be a good deal less percentagewise in the former case, always provided those machines were not bearing nuclear warheads. However, the diffusion and absorption of lasered light by the atmosphere is so marked in dense air as to preclude the effective engagement of low fliers over any very extended range. Yet with interception close-in, a still more fundamental contradiction arises. It is that the narrowness of beam which is the supreme attribute of lasering cannot be properly exploited against aerial targets which are transitting very fast and low, and therefore at an extremely high angular velocity in relation to surface-based defences.

The actual parameters of an attempt at further development along these lines would have to be the subject of more technical exploration. But this much can readily be said by way of broad indication. The metallic skin of an airframe together with the outer casing of some sensitive internal structure (cockpit, electronics suite, fuel tank, weapons bay or whatever) might be burned through by the impact of something like 1,000 joules of lasered light on a surface a centimetre square, which is equivalent to one kilowatt acting on such an area for one second. But when a plane is transitting at, say, 720 km/h and if — as one probably must assume — the laser beam has effectively been set on the one bearing, that plane will pass through the centimetric cross-section of the beam in one part in 20,000 of a second. Accordingly, the flux of lasered energy would have to be 20 megawatts. So on the basis of a conversion ratio of five per cent, one is calling for energy generation equivalent to what might be produced by a small industrial power station!

Naturally, the real requirement would not have to remain quite as unrealistic as that. In spite of what has been said above, greater beam concentration would be bound to improve matters somewhat, as seen from the defender's standpoint. Pulsing would improve it a great deal. So would a general rise of laser efficiency towards the 35 per cent that some commentators believe eventually to be attainable in certain modes. Arguably, too, the apertures of airborne sensors would prove more fragile than the other components just considered. Even so, it is very difficult to accept some suggestions in the aviation press that surface-to-air lasers, effective across several kilometres, could operate on a power input of the order of one megawatt.

Therefore, prognostic parallels with strategic laser development do point to the conclusion that tactical application would indeed be the more daunting. Accordingly, laser anti-aircraft weapons as such are most unlikely to enter service in theatre war this side of the year 2010 (see Chapter 11). Moreover, any systems available then will be static or seaborne surface-to-air ones, only suitable for the point defence of high value sites and ships.[22]

Regardless of this remote and esoteric prospect, however, the indications are that the rates of attrition characteristically inflicted by surface action against monoplanes conducting offensive sorties will soon start to rise more sharply. This will mainly be because, in more and more situations, a critical threshold will be passed through. That is to say, the time taken to engage a transitting aircraft will fall below the time for which it is exposed to fire. To which one might add that the shift may be most dramatic above surfaces that can be described as flat and unadorned; the open sea and those landscapes that most closely resemble it.

Interdiction

The curtailment of enemy reinforcements and supplies has been identified as a salient role of air power ever since the days of F. W. Lanchester and the early H. G. Wells. These last ten years, however, this theme has waxed more prominent again, in central Europe and also elsewhere. One weighty Communist response has been all the concrete hardening (of radar sites, stores, hangars, submarine pens and so on) lately undertaken across North Korea.[23]

Not that the new look at interdiction has been untrammelled. On the contrary, it has engendered a widespread controversy about the relevant allocation of resources. Yet this debate revolves around subsidiary judgements and assumptions that must themselves be uncertain in an environment so fluid, both tactically and geopolitically. What compounds uncertainty in the case of interdiction is the very fact that the impact sought on the main land battle is an indirect one.

Still, a review of past interdiction campaigns may be rendered easier and more instructive by this detachment from the battle zone proper. In his cogent 1936 analysis of aerial events around the trenches of the Western Front, the late Sir John Slessor concluded that 'Attack on supply on the roads forward of railheads will not normally be worthwhile; once supplies of ammunition reach the lorries of the maintenance companies or their equivalent, they become too scattered to constitute a profitable objective.'[24]

His insistence that rail links are much the more vulnerable remains pertinent in the light of a still firm conviction in Supreme

Headquarters Allied Powers Europe (SHAPE) that a strategic reinforcement of the Soviet troops in Germany would be largely railborne. Also, his view is well corroborated by a famous example of successful interdiction: the systematic offensive against the railways of northern France in the early summer of 1944:

> Two or three months of bombing . . . paralysed the economic life of France.
> While the bomber offensive did not affect German troop movements as dramatically . . . highly significant declines had none the less occurred before D-Day in the overall volume of military traffic including building materials for the north coast anti-invasion defences and for German V-weapon installations between Amiens and the Pas-de-Calais . . . During the first fortnight after the landings on the Normandy coast, even troop movements had dwindled to a trickle in the north and west regions.[25]

This was in spite of elaborate constraints having been built into the targeting plan in order to limit civilian casualties.

Correspondingly, the view has been expressed that similar co-ordination of air interdiction might have forced a precipitate German evacuation of Italy around that same time.[26] Indeed, Colonel Trevor Dupuy, a fervent apostle of the statistical analysis of war, has claimed that, even as events transpired, the campaign in Italy indicates that 'sustained interdiction yields six times as much result as a close support effort', on a *pro rata* basis. Here his definition of 'interdiction' is attacks on ground targets (other than airfields, artillery sites and industrial plants) not less than 10,000 m beyond the front line.[27]

Paradoxically, however, it is not hard to imagine a withdrawal beyond the Alpine ridge line of the 30 German divisions then committed south of it having worked to the positive advantage of the Third Reich. The frontal narrowness and tortuous topography that could have made it very difficult to sustain so many troops in peninsular Italy in the face of more determined interdiction also cast doubt on the wisdom of bothering to in any case. Thickening up the garrisons along the coast of France might have been a better alternative, especially in view of the constraints being imposed by Allied air power on more mobile resistance to the Second Front.

Still, the question remains as to how easy it is to interfere with major formations using roads or tracks to redeploy. Certainly, the triumphs of air in World War II include the terrible savaging of the Wehrmacht effected by Allied fighter-bombers (abetted by field artillery) as a partly horse-drawn German army struggled to extricate itself from a near envelopment at Falaise in the final phase of the Normandy campaign. Yet among its more comprehensive failures had been the quasi-interdiction campaign waged against the Axis armies in retreat after Alamein down the zone of rough roads and ill-defined tracks extending back from Egypt to Tunisia. As General Sir Freddie de Guingand, the 8th Army's Chief of Staff, has commented, 'we had visions of the retreat being turned into a complete rout . . . in the event, the results were most disappointing'.[28] This especially applied in regard to Rommel's well-mechanised Afrika Korps.

A decade later, over two years of full-scale interdiction by aircraft backed by naval guns was to be carried out by the UN in Korea: a hilly and populous peninsula remarkedly similar to Apennine Italy. The operational outcome is still debatable in spite of the targeting policy having broadened towards all-out war (see Chapter 1). Thus recourse by the USAF, in the midsummer of 1951, to what was termed Operation Strangle did not effect the isolation of the large Communist armies in the forward zone. Never mind that, even before this particular phase was initiated, movements south along the North Korean railroads are said to have been reduced from 12,000 to 2,000 tons per day. No matter either that heavy suffering was inflicted on the civil population by the very wide casting of the targeting plans or that precision was considerably sacrificed in order to keep aircraft losses down. A major reason for incomplete success was the adeptness shown by the Communist side at railroad repair; the diversion of traffic to roads and pathways; and the dispersal and concealment of stockpiles.

How far then did Operation Strangle achieve the more modest aim of curbing the ability of the adversary forces to wage offensive war? During those last two years of the Korean conflict, the strongest inhibitions against big land offensives by either side were surely (a) the peace negotiations at Panmunjom; and (b) the extraordinary concentrations of men across the narrow waist of the peninsula. Even the United Nations was deploying one division for every eight or ten miles of the 140-mile front, while their adversaries were managing virtually one line division every two miles. What is more, the latter were also able to supplement their otherwise modest scales of field artillery with specialist artillery divisions, as circumstances determined.

Therefore, the Communist ability to maintain their density of troops to space can only be seen as a rather negative commentary on the endless interdiction strikes. The same goes for the reported ability of the Chinese and North Korean armies to hold in the field between 30 and 60 days of supplies, a level high by any standards but especially by those of immobile war.[29] So does it go for the large number of strong local attacks the Communists launched, as witness the ferocious battles for 'Pork Chop Hill' towards the end of hostilities. Nevertheless, a masterly overview of this campaign included the tentative judgement that 'but for the interdiction campaign, Chinese ground activity might have been much heavier during the stalemate war'.[30] What air interdiction may particularly have curbed is the exercise by the enemy of their field artillery.

The next massive interdiction campaign was that waged by the USAF, from 1965 onwards, against the 'Ho Chi Minh trail' from North Vietnam through Laos to South Vietnam. Duly, the said corridor was turned into a veritable test site for the ultra-electronic battlefields of the future:

> To detect movement of trucks or personnel along the trail, sensors are implanted on the ground or suspended in the foliage by air drop. Aircraft overhead receive electronic messages from them and relay the information to a central computer control station.[31]

However, this exotic technology had too much to cope with from the thorn scrub and rain forest in which the area abounds, as well as in regard to the facility with which lorries, carts and porters could disperse across innumerable rough tracks. Besides which, the weight of supplies needed to sustain the campaign in the South was a modest 15 to 75 tons a day. Therefore, even accepting without reserve the bold claim that, by 1970 or thereabouts, 70–85 per cent of all materials dispatched by Hanoi were being destroyed *en route*, one might still doubt whether these interdiction strikes ever did much to limit the military pressure exerted by the Communists in the South.[32]

Nevertheless, all the historical evidence indicates that, except where geography or troop densities are entirely unfavourable, sustained interdiction can badly impede the progress from the deep rear to the front of major reinforcement units. After all, formations of the People's Army of Vietnam regularly took ten to twelve weeks to move down the Ho Chi Minh trail. Similarly, during the Korean War a division of the People's Liberation Army might have taken three or four months to move from the Yalu to the front. Both assessments imply average progress of several miles a day. Likewise, a Wehrmacht division crossing northern France in July 1944 would be lucky to exceed twelve miles a day. Nor should one forget that, when units are slowed down too drastically, they may also incur heavy casualties. In any case, their cohesion and morale are bound to sag. These then are the grounds on which air strategists aver that the sheer dislocation caused to an army on the offensive may multiply several times the impact of interdiction. As a Luftwaffe assessment has put one aspect of this, a delay of one hour along one axial road axis, typically retards by three the advance in that sector.[33]

Furthermore, the mere threat of air attack can oblige supply columns so to spread themselves that the carrying capacity of axial highways is reduced by at least a half.[34] Alternatively, aerial ECM — maybe conducted from rotorcraft — can so hamper the wireless control of mechanised columns as to make them close up enough to render air strikes all too punishing.[35]

However, certain issues cannot be avoided apropos of the future. The most basic is what emphasis ought generally to be placed on interdiction as opposed to Close Air Support. Should any credence be allowed the verdict Colonel Dupuy reached in the light of the Italian campaign in World War II?

The first truism to note is that the warload penalties incurred through striking deep will usually be severe. If intrusion has to be fast and low, indeed, the constraints thus operative may quickly prove absolute, not least with manned aircraft. Even planes as sizeable and adaptable as the Tornado and Jaguar have radii of action of 750 miles (with a good warload borne externally) in a 'hi-lo-hi' flight profile, yet only one of 450 if it must be 'lo-lo-lo'. Another big problem arises in respect of the acquisition of targets such as bridges. After all, the Forward Air Controller (FAC) cannot be contacted by a deep striking pilot, except in those special cases where he may infiltrate by parachute or whatever. What is to be hoped is that the enthusiasm for interdiction resurgent within NATO does not owe too much to a dismissive attitude towards the sort of air-land integration the FAC represents.[36]

At all events, the axiom that an extensive air threat obliges the opposition to divert what easily become disproportionate resources to anti-aircraft defence can weigh heavily in favour of deep interdiction. This is mainly because the objectives to be defended will tend to be well spread out. But it is also because whatever dual-capability (in the sense of counter-surface as well as anti-air) an anti-aircraft weapon may possess cannot be exploited if it is located too far to the rear.

All the same, most armies nowadays are inherently less susceptible to air interdiction than was the case in 1944, say. Tyre-mounted automobiles have better off-road performance, while horse-drawn carts are virtually a thing of the past. Many more tracked vehicles have entered service, notably with the infantry and artillery. Vehicular adaptation to swimming and wading has further been refined, though entry into and egress from a stretch of water may still present problems. Meanwhile, the technology of high speed bridging has advanced appreciably, representative examples being that a 50 m 'dry support' bridge can be built in 25 minutes while a pontoon bridge can be laid at a speed of a good six metres a minute. The rapid repairs of the Suez bridges in 1973 confirm this aspect of adaptive technology. A measure of its importance is an analysis of the central European theatre which suggests that a 20 m bridge is typically required every 20–25 km; a 100 m one every 50 km; and one of up to 150 m every 100–150 km.[37]

No doubt those who extol the particular merits of Battlefield Interdiction — that directed against the immediate rear of the battle area itself — are impelled in part by an apprehension that a future main engagement could be so dynamic and shortlived that air strikes against the logistical infrastructure in the adversary's deep rear would little affect its outcome. However, they also insist that, in most environments, the matrix for the immediate resupply of engaged troops has many highly visible nodes of vulnerability: 'Ammunition is unloaded and reloaded at regiment, division and Army dumps. Wherever this process occurs, a target is created . . . Fuel supply . . . also is beset by transfer and storage problems'.[38]

Another theme to examine is this. Heretofore the received wisdom about air interdiction has been that it is the last 10 per cent of effort that 'gets the real pay-off'. In other words, there are, in the final analysis, sharply increasing returns. In the future, however, it may more normally be a matter of returns at the margin diminishing throughout. A low-key campaign may harass the adversary disproportionately. Yet a substantially greater one may not make that much extra difference. For instance, it will rarely make sense under modern conditions to seek to isolate the battlefield absolutely by means of 'a virtually impassable zone'.

Still, a generalisation so abstract must be subject to some qualification. Much may depend on whether an interdiction campaign is combined with one against military airfields. So may it on whether it might be broadened into an attack on the adversary's political and economic order. And within the sphere of interdiction as strictly defined, could attacks on the command-and-control structure yield singular benefits? How exposed will this part of his infrastructure be to electronic and optical surveillance? How fragile will it prove in the face of modern ordnance? How dependent is the adversary on his control facilities? Can his leadership cadre be suitably cowed or distracted by strikes deep into it, much as Hitler appears to have been by Anglo-American strategic bombing? But are there not risks in degrading the said structure overmuch, bearing in mind the interest our side may have in his being able properly to restrain his own nuclear or chemical release? Too much of the commentary on Battlefield Interdiction in particular fails to address these issues, thereby becoming itself a rehash of modish doctrine.

Tactical Reconnaissance

Close Air Support and, even more particularly, interdiction directly depend on reconnaissance, both before and after. But the ability of aerospace power to 'see the other side of the hill' is of immense importance for the modern military art as a whole. Thus during World War II, 60 per cent of the background intelligence about opposing forces in the European theatre and 90 per cent in the Pacific derived — say US sources — from the interpretation of aerial photographs. During the highly dynamic invasion of Manchuria in August 1945, 30 per cent of all Red Air Force sorties were for reconnaissance.[39] Today, squadrons dedicated to this task exist in many air forces. Western estimates are that one in every eight of the Warsaw Pact's first-line warplanes is earmarked thus. In the Luftwaffe the fraction is as much as four in 26. Even in Switzerland, where little scope exists for open mobile warfare, one out of the 19 listed flies the reconnaissance version of the Mirage 3. What is more, several aerospace industries have evolved (and many air arms employ) reconnaissance pods. These usually have a mix of optical and infra-red sensors, and can be attached to a variety of tactical aircraft as occasion arises.

Many of the aircraft specifically dedicated to tactical surveillance over land and narrow seas are strike/fighters or light bombers adapted in either their prime or their obsolescence. But there are other lines of development. Long-Range Maritime Patrol (LRMP) planes blend the faculties of strategic surveillance, tactical warning and tactical engagement. Then again, various army or army co-operation light monoplanes have been designed with local reconnaissance mainly in mind. As much is always betokened by rough-field capabilities, wide downward vision and a general solidity. Each of these aspects currently finds most explicit expression in the Optica: a three-seat British monoplane with an all-up weight of 1.2 tonnes and an ability to use 800 foot airstrips.

Meanwhile, the reconnaissance variant of the USSR's MiG-25 interceptor (35 tonnes all-up weight) appears to achieve Mach 3.2 in high altitude bursts. Thus does it revive with a flourish the principle, well applied in World War II by the Spitfire and Mosquito, of survival through height and speed. Evidently, too, this principle informed the development (20 to 25 years ago) of the Lockheed SR-71 Blackbird, a machine of 70 tonnes loaded weight. It may travel at 90,000 feet for some 3,000 miles and at speeds of up to Mach 3.5.

On the other hand, though the TR-1 derivative of the famous Lockheed U-2 travels as high, its speed barely exceeds Mach 0.65. Two tons of sensory equipment include a Sideways-Looking Airborne Radar (SLAR) which sends data direct to the ground and which is credited with slant ranges of up to 300 miles. A Precision Emitter-Location System is among its other systems. Therefore, two or more TR-1s may together fix a hostile emitter through triangulation.[40]

The TR-1 came first into service in 1981 to offset the cancellation in 1977 of work on Compass Cope, an RPV of 13,000 lb all-up weight designed to loiter for 30 hours at 80,000 ft and then land in any weather. The radar image of the latter had been quite small. Yet it had over 100 cubic feet of avionic circuitry plus antenna space of similar proportions. An anticipatory comparison was made between this aerial vehicle patrolling 60,000 feet up along the inter-German border and an RF-4E Phantom doing likewise at 30,000 ft, each machine being tasked to monitor Warsaw Pact overground deployments to a depth of 100 km beyond the said divide. In the lowland sector from the Baltic Coast to Magdeburg, the Compass Cope was credited with a 99 per cent coverage as against 60 per cent for the Phantom. From Magdeburg to where Czechoslovakia abuts on Austria, the respective estimates were 94 per cent and 31 per cent.[41] Thus the former was reckoned to perform more consistently over varied terrain than could the latter.

Allegedly, the cancellation of the Compass Cope in 1977 (after four years of flying trials) owed more to matters political and circumstantial than to those technical and intrinsic.[42] At all events, there can be little doubt that the kind of capability sought through this system will be available in the West within a decade. As a genre, indeed, the Remotely Piloted Vehicle (RPV) and the drone are destined to figure prominently in aerial reconnaissance, not least because many models can adapt to other roles. Several thousand such devices were launched by the USA during the Vietnam war. One claim is that '85 per cent of the photos taken to assess bomb damage . . . were brought home by these automated craft'.[43]

At what may be termed the further end of the panoply of unmanned surveillance, there is reconnaissance from near space: the only form of overhead surveillance permissible in peacetime under international law. Within the US programme at any rate, the ground resolution obtained from orbit by automatic photography (probably when enhanced by the computerised integration of several runs over the area specified) is now six inches or thereabouts.[44] In other words, a patch of distinctive colour that wide would appear as a dot on a film. Therefore, objects some 10–15 ft across (e.g. vehicles on roads) will often be identified — at least in outline. Such acuity is actually rather better than was first achieved by the U-2s some 30 years ago. Moreover, this performance is sure to improve further, especially through better computer resolution of successive overlays. Never mind that satellites in low orbit pass over a given location only once or twice each daylight phase.

Already, too, this technology is supplemented by automatic listening for radio and radar emissions and also by infra-red scanning, especially of concentrated heat sources. Now interest is reviving in the USA in the possibility (explored prematurely in the early 1960s) of tracking newly-launched missiles by the infra-red signatures of their exhausts.[45]

The Classic Missions   145

None the less, satellites as a genre do have their drawbacks. Among them is the vulnerability to interception inherent in a discrete object moving in isolation in an orbit largely or entirely predictable and well above the tangible atmosphere. Nor is it at all easy to make the sensory cells on board at once sufficiently tough and suitably responsive. Sometimes, the most tempting remedy might lie in the forceful suppression on their launching sites of the anti-satellite weapons. Yet that might be tantamount to non-nuclear strategic war.

To some extent, the routinised inflexibility so characteristic of this mode of surveillance can be offset by the launching of 'bursts' of satellites to cover a particular regional crisis. This stratagem figured, indeed, in the Soviet response to the Lebanon and Falklands campaigns in 1982. Likewise, the USAF has been examining the feasibility of a 'Space-plane' that, at a few minutes' notice, could pitch 10 tons of satellites directly into low orbit. It might make two sorties a day from any large airfield.[46]

Meanwhile, the rate and quality of data transmission remains a weakness. Telemetry may degrade resolution to an unacceptable extent; and a full read-out to a given location may be hindered by the way most satellites completely orbit the whole Earth roughly every hour-and-a-half. The ejection of films and tapes by capsule is a long-established alternative but (a) recovery and processing can take hours; and (b) the general practice has been to have a capsule eject only after a week or two.

What is more, photography by starlight or moonlight is still in its infancy. Nor can satellites circumvent cloud cover. When skies are clear, orbiting satellites can register atmospheric profiles of temperature and humidity which are remarkably definitive, except perhaps at the lowest levels. In fact, this is a prime reason why it is today possible — in central Europe, say — to make general weather forecasts that are 60 per cent valid ten days ahead.[47] Yet that very facility might work against the use of satellites in war. This is because offensive action can now be planned to coincide with sustained cloudiness (see Chapter 13).

Herein lies an argument in favour of reliance on low-flying aerial vehicles that is not to be lightly dismissed. Take, for instance, the averaged figures for West Germany. Cloud descending to less than 1,000 ft above the surface is present at any particular place on 18 per cent of occasions. But cloud confined to 1,000 ft or above is present on a further 56 per cent.[48]

Ostensibly then, a combination of aerial and near-space observation does uncover those beneath in the comprehensive way the classic air theorists dreamed of. Moreover this capability is continually being extended by technical advance: computer enhancement, passive infra-red, low-light television and so on. Hence some authors talk of a 'transparency revolution' whereby everything becomes observable. Supposedly this is to the singular advantage of the defence.[49]

Clearly, the last stage of the argument rests on the Liddell Hart fallacy. This apart, however, one must not exaggerate the completeness of the transparency effect. These last 40 years or so have seen dramatic examples of either the camouflage or the false representation of large formations of troops, not least on desert landscapes. Before the battle of El Alamein in 1942, the British had liberal recourse to disguised tank parks and fake pipelines, supply dumps, assembly areas, etc. Hundreds of dummy vehicles were erected in a deception directed mainly against air reconnaissance. This unsuspected ruse seems to have inhibited German responses to the main attack.[50] Similarly, before the Israelis attacked in the Sinai in 1967, they moved to a brigade just opposite El Kuntilla. Then dummy tanks, masked by suitably inadequate camouflage nets, were fed in to form two phantom brigades alongside. True, the relatively uniform colouration of the desert favours such stratagems, as does the relative stability of its weather. But, of course, a lack of vegetation could work strongly the other way.

Similarly, even maritime geography affords a good enough milieu for this purpose. The two world wars saw considerable recourse at sea or in harbour to either disguised or dummy naval vessels, not to mention routine ship camouflage and concealment through smoke. A related artifice is the comprehensive shrouding of vessels at anchor. Sweden is possessed of a broken and indented coastline, along which hard rock often plunges almost sheer to and below the water. So she has brought this particular art near to perfection. She has done so principally by means of camouflage nets that blend visually with the pre-selected mooring places and which may also reduce by an order of magnitude a vessel's radar image, this over a wide range of frequencies. Ships as large as destroyers can be hidden thus; and to cover a Fast Patrol Boat (FPB), say, will normally take but twenty minutes.

Dummy airfields with their due complement of dummy aircraft also figured in World War II, the final Soviet offensive in the Far

East again being a notable example. Today, much more sophisticated dummy planes, intended to fool a diversity of sensors, are available in various countries. Erecting one may take two man-hours.

An indication of Soviet prowess at frustrating radar surveillance, in particular, is as follows: 'In addition to natural cover, timber brushwood, metallic nets and angle reflectors are used by Soviet forces for radar camouflage. Mock-ups of military equipment can also be used as anti-radar reflectors.'[51]

Irrespective of deception, however, the provision of aerial reconnaissance data in the right form at something like the right time and place is far from assured, particularly when judged in relation to the rising tempo of war. The processing of camera film is liable to take 45 minutes; and integrating optical evidence with other kinds may take a further 30 minutes. Nor should we overlook the time a plane or RPV will take to return to base. Video-telemetry or voice communication can save precious minutes but may well depend on relay via another plane or a satellite and may be too susceptible to interference. Besides, the data thus obtained may be hard to marry with other 'real-time' evidence.

So how do these several considerations relate to the evolving balance between manned vehicles and unmanned ones? How much substance will remain in the longstanding and oft-repeated claim that, in this aspect of aviation as in others, the former are inherently more flexible? Granted, reconnaissance satellites are peculiarly inflexible systems by any reckoning. But may not RPVs and reconnaissance drones gain a lot from their ability to fly, and indeed to loiter, far above or low above the ground even within fiercely hostile environments? Aircraft are much less flexible in respect of altitude in particular. This is because they may be more prone to being shot down and yet cannot be regarded as so readily expendable.

Nor may the crew members of an aircraft be as reactive as automated sensors are, especially in the context of flying fast and low. Here the rub is visual acuity, a limiting factor being the angular velocity at which a pilot or observer swings past objects on the surface. At 800 mph, say, ground 200 ft immediately below is 'racing by' at 275 degrees a second, over three times the 'threshold of visual perception' value identified in Chapter 3. True, tanks or whatever on rising ground 1,000 yards or more away would still not present that particular problem. On the other hand, they might have been camouflaged or otherwise hidden. All of which lends credence to the oft-heard view[52] that, the Israelis excepted, Western nations have not taken due advantage of the art of unmanned reconnaissance as extended by technical progress through the 1970s.

Nor is it just a question of improvements in surveillance technique. May not a shift of emphasis to 'extended front' dynamic war (see Chapter 5) afford more scope for 'mini-RPVs' that may weigh less than 100 kg and have a radius of action of perhaps 50 km. This is because it will be more of a brigadier's or colonel's war, figuratively speaking, and less of a general's one. In other words, it will be shaped more at brigade or even battalion level and less at that of division or army. Let us recall a US Army precept about requisite depth of reconnaissance. Generals have to see 150 km beyond the Forward Edge of the Battle Area (FEBA) though colonels only 50 km.[53]

Manifestly, dramatic differences in comparative cost (see Chapter 12) strengthen this case for the RPV. Yet it seems less conclusively made when the time lapse is also considered. Thus even intelligence sought at army or divisional level may lose much of its operational value unless fed through within the hour. At brigade headquarters, intelligence updates on the enemy's overall dispositions ought to come to hand within 20 minutes. What is more, data to be used for the indirect aiming of fire against mobile targets may well be needed within a minute. Unfortunately, however, institutional and other human impediments may make such deadlines hard to meet. Allegedly, indeed, US forces in Vietnam found that it sometimes took several days for objectives discovered in reconnaissance sorties to appear in regional targeting plans.[54] Telemetering data from warplanes or manned rotorcraft as it is being obtained can be one answer.

Historically speaking, various examples could be cited of intelligence estimates of opposing strengths proving far less assured or exact than might have been expected, given the then state of art and all the contingent circumstances. A famous instance is the way radio silence and other modes of concealment, much assisted by bad weather, enabled an entire panzer army to lose itself to Allied observation in the fortnight prior to the Ardennes onslaught of December 1944. Hardly less remarkable is the failure of the Israelis to pre-empt the Arab offensive of October 1973 (see Chapter 1). They were unable to read the signs with enough assurance to convince both themselves and the Americans that an onslaught was in the offing.

Several factors contributed to their inhibiting doubts. Among them was the elaborate Arab deception plan. In addition to spreading their assault forces all along the Suez Canal, the Egyptians employed a whole panoply of more specific tricks: positioning tanks and artillery as though for defensive action; hiding bridging equipment in large crates and pits; using underground cables to transmit the more sensitive messages; further obscuring the general forward movement by means of unit rotation; and so on. Yet as their Chief of Staff, Lieutenant-General Saad el Shazli, was to put it, 'The last three days were especially difficult . . . we did not expect the enemy to be taken in as easily as he was'.[55]

Yet human factors notwithstanding, surveillance is now poised to move ahead of concealment and deception, at least in most geographical environments. The passive sensoring of infra-red emissions is one area of development that looks especially promising,[56] not least with reference to nocturnal deployment. What remains much less certain, however, is whether the improving prospects for reconnaissance are matching the new dynamic of land warfare.

Air Defence in Depth

The protection from surveillance and attack of the rearward zones will always include the 'point defence' of key airfields, bridges, headquarters, warships and the like by means of clusters of surface-to-air weapons. Nevertheless, the crux of the matter has always to be where to strike the balance between zonal defence by Surface-to-Air Missiles and that by warplanes. Certain of the arguments either way are elementary. Clearly, only aircraft can inspect aerial intruders, and caution against continued encroachment. On the other hand, a missile, being light in disposable load and driven by very basic mechanics, is far cheaper than a large interceptor plane. Usually, too, it is less liable than the latter to be put out of action before it has left the ground. After all, a SAM launcher is compact and robust in character and is relatively independent of support facilities, its own radar excepted. Besides, even if an interceptor aircraft is on runway alert, it will still take a couple of minutes or so to become airborne.

Nor will it then climb to high altitude as fast as a SAM. From the moment of take-off, even a stripped-down Phantom has taken 35 seconds to reach a height of 3,000 m; 77 to reach 12,000 m; and no less than 230 to reach 25,000 m. A largish SAM, on the other hand, would probably take some 20 seconds to reach 12,000 m, and then maintain much the same rate of progress much higher into the stratosphere than a Phantom or whatever could manage.

In other aspects of responsiveness, however, the kind of SAM here being considered compares less impressively with the warplane. True, it will accelerate forward at 30 to 50 g throughout its 'booster' phase, either in immediate hot pursuit or else to achieve a high enough speed at burn out to sustain a long enough interception track. However, its lateral acceleration (only a minor fraction of the above, even in the denser air at low altitude) may fall to the 2 to 4 g bracket near or in the stratosphere. For one thing, large control surfaces cannot be fitted because of initial cost and aerodynamic drag. For another, there is no disposable energy once burn-out has occurred. Both drawbacks affect it most in rarified air.

Beyond that, precise information about SAMs as a genre is hard to come by. All that can be said is that gains in agility at high altitude are unlikely to match those pending near the ground simply because different factors are critical. With the interceptor or air superiority fighter, on the other hand, up-to-date evidence about an ability to turn and so on is quite adequate for this analysis. On graphs of altitude plotted against Mach number, contours are drawn to indicate the range of speed within which a given lateral acceleration can be sustained at particular altitudes. Thus a present-day fighter might register 9 g close to the surface but only 6 g at 8,000 ft, at speeds close to Mach 0.8 in both cases. What then of its successor a decade or so hence? Even it will not achieve a forward acceleration much in excess of 1 g. However, its lateral attainment should be 8 g (between Mach 0.7 and 0.9) at the medium altitude just cited, and 6 g near Mach 0.9 another 10,000 feet up.[57]

Admittedly, a heavy external warload may sharply reduce such agility. But what can also be remembered is that, other things being equal, the extra speed of an approaching missile will make turning 'inside it' that much easier for the aircraft being pursued. Suppose a missile is travelling half as fast again as its prey, a manned aircraft. Then it has to 'pull' (i.e. exert upon itself) 2.25 times as much lateral acceleration to stay within the same turning circle. In other words, a square law applies and affords the aircraft a chance of evasion, especially if exploited in conjunction with ECM.

Admittedly, some elementary factors work the other way. An approaching SAM may go undetected. Several may approach at once. Any SAM missile should always be able to start to alter direction a second or two quicker than its prey. But suffice for the present to suggest that the most apposite form in which to couch comparisons of this aspect of manoeuvrability is not lateral acceleration *per se*. It is the actual rates of turn in degrees of arc per second of time.

Yet once more, little hard information can be gleaned in respect of SAMs. What can be done, however, is calculate that the exertion on a SAM travelling at Mach 2.5 of a lateral acceleration of 12 g, say, will cause it to turn through a quarter circle at an angular rate of some nine degrees a second. Meanwhile, the evidence apropos tactical warplanes is again more direct and definitive. Through the late 1960s, the turn rates they sustained reached 10 to 12 degrees per second. This was at low altitude and at speeds optimal in this regard, which usually meant well subsonic. However, the sustainable turn rates in new fighter specifications may be as high as 20 degrees per second at sea level; and still be six or so at Mach 1.5 and 36,000 ft. Conceivably, in fact, no less than 30 degrees will be briefly attainable low down.[58] To be more specific, the Luftwaffe submission for the Outline European Staff Target for a new multinational fighter envisaged an 'instant' rate of turn of 25 degrees per second and a 'continuous' one of 14 at low altitudes and a transonic speed.[59] The corresponding turning circles are 1,560 m and 2,800 m in diameter.

Needless to say, these projected improvements reflect progress in propulsion and in airframe design. As implied above, the manoeuvrability of a given machine is largely a function of the size and responsiveness of its lifting and control surfaces and of the disposable propulsive energy. Regarding the latter, Specific Excess Power is a proportional indication of the energy available over and beyond what is needed to sustain flight straight and level. It is arrived at by multiplying velocity and the excess of thrust over drag; and then dividing the product thereof by that of mass multiplied by g. Suffice to note that banking at a steady rate of 25 degrees a second on a circle 1,560 m across would be produced by a centripetal (i.e the opposite of centrifugal) acceleration of 15.0 g.

152 *The Classic Missions*

A major reason why that steadiness could never be maintained for the seven or eight seconds required to complete a half turn, say, is that no aircrew could stand such a g-stress more than momentarily.

The conclusion which emerges is this. Even when combined with advanced ECM, improvements in the agility of monoplane aircraft are not going to alter beyond recognition their chances of survival at high altitude in the face of a manifold SAM threat. All else apart, the comparison of basic cost is much too favourable to the SAM (see Chapter 12). Two broader inferences follow. The one is that strike aircraft are not going to revert massively to conducting their sorties at medium and high altitude. The other is that the interception of these machines will remain very considerably the prerogative of SAMs.

Still, the dialectic does not end there. Among the positive attributes of the monoplane is that, as a big and high-flying air-breathing vehicle, it always has quite a wide radius of action. In contrast, the largish Bomarc B, introduced into the air defence of North America some 20 years ago, was utterly exceptional as a missile able to intercept 650 km out, at least when fitted with a nuclear warhead. It could do so partly by virtue of its sizeable lifting surface. More particularly, however, its main motor was not the usual rocket but a 'ram-jet'. This is a jet engine in which the turbocompressor has been dispensed with because the spontaneous inrush of air in fast flight can be relied on to ensure a due mix of oxygen and fuel. Still, the radius cited only begins to compare with, let us say, the F-15 Eagle's ability to intercept well over 1,500 km from its parent field, even without refuelling in flight. Besides, a much more modest 80–120 km is the representative spread of slant ranges to date, even with SAMs intended for strategic air defence.

Moreover, this situation cannot be attributed simply to an obligation to keep such systems tolerably cheap. Nor can it to the fact that a rocket has to carry oxidant as well as fuel. Another constraint is the ineluctable fact that radar beams diverge, thereby making both target-tracking and 'beam-riding' progressively less sure over distance. A missile which must therefore be corrected too abruptly will be prone to unstable oscillation in flight. Still, either radio command or self-homing may help to offset this effect. Bomarc was guided by both, in fact.

With the duration and autonomy of the manned aircraft goes the ability to loiter on Combat Air Patrol (CAP). Several of the major air forces of World War II regularly made such a present of power.

They did so in pursuance either of air defence or else highly reactive Close Air Support (CAS). Nevertheless, CAP has always had its critics. To Lord Tedder, indeed, the 'air umbrella' mode of defence was 'both the most extravagant and normally the least effective; a method which, if persisted in, is a sure way of losing any degree of air superiority you may have'.[60] All the same, there will sometimes be commitments (not least the protection of ships at sea) which leave little alternative. Lately, the provision of deep ocean cover has been greatly facilitated by in-flight refuelling by machines adapted as aerial tankers (e.g. the RAF's VC-10s, Victors and Tristars).[61]

These days, too, the defensive employment of fighter aircraft (whether on CAP or from 'scrambling' off the runways) is often to high altitudes. One attraction is that the silhouette of virtually any aircraft looks a great deal bigger from overhead (or, for that matter, underneath) than it does from dead ahead or dead astern. Take, for example, the MiG-21. The image it projects to a viewer directly in front is equivalent to that from a perfect reflector, three square metres in area and set flatly abreast of the field of view. But to someone on a slant bearing of 60 degrees above or below, 23 sq.m are thus projected; and to somebody directly overhead or beneath, 27 sq.m.[62]

Still, the most basic geometrical truism is the dramatic way the potential field of vision increases with altitude. Nor do the figures given in Table 7.1 tell us all. Thus the maximum intervisibility attainable between two aeroplanes in flight will be the sum of the respective slant ranges to the horizon. In practice, however, these perspectives are best exploited via aircraft designed for Airborne Early Warning (AEW) (see below, p. 198).

Yet even with such aid, fighters patrolling at altitude rarely respond to threats as readily as geometry might lead one to expect. So it is important to ask why. My own experience of simulated air combat near or in the stratosphere has been one sortie in the observer's seat of a Royal Navy Venom in 1958. Still, that was enough to convince me how much harder visual acquisition is up there. A pilot does tend, when background features are lacking, to focus his gaze much too close. Moreover, reduced scattering in a thinner atmosphere makes the background light at, say, 30,000 ft a mere tenth of what it is at sea level. True, trails of water droplets or smoke may still betray another plane from quite a long way off. However, condensation is liable to persist only within narrow bands of altitude, this persistence depending on a subtle balance of low ambient temperature and high relative humidity. Nor is exhaust smoke often a giveaway. The MiG-21, for instance, hardly ever leaves a trace.

Visual recognition is likewise constrained. Without special sights, warplanes can rarely be identified beyond several kilometres or helicopters beyond one or two. Furthermore, the visual method of range guestimation remains the one adopted in World War II. By means of a graticule, you compare the angular width of the target aircraft with its known wing span — assuming its type has been identified. Given a steady bearing and a moderate closing speed, this can be accurate to within 100 yards. Otherwise its performance is poor. But with radar ranging, the standard error is more consistently cut to ten yards or thereabouts; and with laser ranging further again. In addition to which, these last two methods lend themselves to computer extrapolation. Meanwhile, electro-optical aids to human observation are set to come more into play.

However, the limitations of electronic Identification Friend or Foe remain all too evident, especially in ECM environments. But just how much these will matter in themselves will depend on the total circumstances of the air battle: its spatial limits, density and general coherence. Some encounters can be managed so coherently that the allegiance of aircraft can confidently be inferred from their observed movements. Others are chaos without good IFF.

Next, one ought to address the shortcomings, as a means of surveillance, of radar borne in fighter aircraft. One requirement is to obtain a tolerably concentrated beam from a narrow scanner. Therefore, quite a number of sets operate on frequencies as high as 10 to 20 gigahertz; the J-band in terms of the new nomenclature cited in Appendix A. Yet in this part of the spectrum even thin upper cloud or light low drizzle starts to interfere with the emissions. Then again, the main radar in a fighter typically emits a peak power an order of magnitude less than the couple of megawatts or so usually achieved by master radars on the ground. The 600 kilowatts registered on the MiG-25 is very exceptional.

Another problem is that, in air-to-air combat, the radar requirement assumes two distinct forms: search and acquisition. The former may best be effected by transmission on not too high a frequency because this may help ensure adequate beam divergence. Pulse repetition rates are best kept fairly low, too, the better to gauge the range of initial contact from re-echo arrival time. In the fire-control mode, on the other hand, the accent has to be on full and precise data from a pre-determined quarter across a limited range. Therefore, preferences shift towards higher frequencies and good beam focus coupled with narrow pulse width and rapid pulse repetition.[63]

In some machines — for example, the Phantom — this conflict of priorities is resolved by installing two radars. These days, however, it may be thought best to use a single set able (albeit with penalties in weight and efficiency) to switch between alternative pulse repetition rates, azimuth and elevation scans, and even wavelength frequencies. In fact, the F-16's main radar is credited with six distinct modes, with operating ranges which vary from 2 to 80 miles. Likewise, the Mirage 2000 (now entering service with the Air Defence Command of the French Air Force) is said to have a maximum surveillance range of 65 miles. For other interceptor or multi-role aircraft, rather lower figures are cited. So very often the extent of the effective scan will fall well short of the distance to the geometrical horizon. Nor can one feel confident that radar performance in this setting yet matches adequately the potentialities of some of the larger air-to-air missiles, especially when the ECM risk is allowed for. Already, a Phoenix fired from an F-14 Tomcat has intercepted a drone across a distance of 110 miles. Six such missiles (plus other ordnance) can be borne by a Tomcat; and may be dispatched in very quick succession.

One subtle yet important theme is 'look-down, shoot-down': the presumed ability of the latest generation of interceptors regularly to engage, across extended slant ranges, intruders travelling at low altitude. The main impediment is that, within several hundred feet of the Earth's surface, the radar reverberation therefrom (what is colloquially known as 'clutter') tends strongly to mask the intruders. True, this effect may be counteracted by doppler discrimination between objects moving and a stationary background (see Appendix A). However, since no pulse can be stably transmitted on an absolutely constant wavelength, the doppler shift in apparent wavelength will be more difficult to discern when the received echo is weak. So at long-range, low-flying aircraft will usually be obscured unless the velocity vector they present to a radar set is several hundred miles per hour.

An indication of what this may currently mean in practice has been afforded as follows. An F-15 travelling at Mach 0.9 and 20,000 ft would be able to employ an AIM-7F missile against, say, a MiG-23 at the same altitude and 27 nautical miles away. When the F-15 is at 15,000 ft but the MiG-23 is no higher than 1,000 ft, this mileage is reduced to 23. Meanwhile, a very similar capability is conceded the MiG-23 and MiG-25, respectively armed with AA-7 and AA-9 missiles. Thus a two-seat advanced version of the MiG-25 has been observed engaging drones at 1,000 ft from an altitude of 20,000 ft. Once, in fact, the drone was below 200 ft at a slant range of 12 miles.[64]

In all probability, this last kill was over the sea where clutter is less, at least if the surface is fairly calm. But whatever assumption is made on that score, it can hardly be said that this or the few other reports that have appeared enable us to draw clear-cut conclusions about the present efficacy of this mode of interception within a few hundred feet of the surface, especially the land surface. For one thing, it is not clear how sharply the interception prospects diminish as an intruder underflies an interceptor. The point is that the steeper the angle of attack becomes, the less the velocity vector presented by the intruder and also the greater the risk that a homing missile will plunge straight into the land or sea. Still, there is little doubt that a worthwhile capability for 'look-down, shoot-down' is already possessed by both the West and the USSR. Also, some modern planes fire missiles at adversaries flying scores of thousands of feet higher than they are themselves.

Meanwhile, the smaller of the air-to-air missiles offer manned interceptors something of the extra agility they may need at close range. They do so because they are so configured and since the lower speeds they attain allow fairly tight turning circles. None the less, nearly all the latest fighter designs have reaffirmed the value of the cannon, the MiG-25 being a conspicuous exception. Thus among the designs that do sport a cannon are the F-14 Tomcat and F-1 Hornet; the Mirage F1 and 2000; the Tornado; and the MiGs-23 and -27. Obviously cannon fire is the better way to support inspection with warning shots. It is also more eligible for ground-attack. Still, the crux of the matter has to be where the balance is likely to be struck in air-to-air warfare as between stand-off engagements across a well defined zone of separation and encounters at close quarters. Only the latter puts cannon in with a real chance.

On first examination, the former interpretation, at least if not too idealised, looks the more valid. At all events, there has been a swing away from the dogfight image, certainly as compared with its romantic heyday in the aerial sagas of World War I. To quote one historian:

> In the intensified air fighting towards the end of the Second World War, it is significant that the most successful pilots generally avoided the classic dogfight. Instead, they developed specialised attack patterns, usually a fast pass from above taking opponents by surprise, a close-in burst, then disengagement by diving or climbing out of a speed advantage.[65]

Likewise, Soviet analyses of the Great Patriotic War show that two-thirds of all the Luftwaffe planes destroyed in the air were shot down in a single pass attack. Even in 1914–18, indeed, the more prolonged of the aerial duels on the Western Front seemed likelier to end in stalemate than did the short and sharp encounters.

No doubt, too, some would support the case here put by a process of reasoning analogous to that unfolded by Basil Liddell Hart about the ratio between troops and linear space in ground warfare (see Chapter 5). Thus the fact that planes like the Tomcat can use missile-fire to engage at long range several intruders at once does encourage the expectation that a given number of fighters on CAP could cover a much wider sector than ever before. Yet here, too, such ratiocination may be too glib. All else apart, cruising time may constrain CAP, especially in situations where refuelling in flight is impracticable. So, even more critically, may the expenditure of ordnance. Once their larger missiles had been released, aircraft would have no choice to make except that between complete disengagement and coming to much closer quarters. Also, attrition by the main SAM belts for anti-aircraft defence in depth might contribute to a breakdown of so measured a confrontation.

What is more, the history of the last two decades suggests that something approaching a classic dogfight may sometimes recur quite readily. This happened frequently over North Vietnam, partly because of the inhibitions on long-range encounters induced by deficiencies in IFF and partly because the North Vietnamese sought thereby (a) to exploit the then exceptional facility with which their MiG jets turned, though (b) to lessen their dependence on rather backward electronics. To which one might just add that authentic dogfights have figured in the Arab–Israel conflict on various occasions.

Throughout we should bear in mind the oblique but ongoing

debate as between the Western Europeans and the Americans. Might not a future air battle conform to no coherent pattern of any sort? Might it not instead be more of a mêlée, a chaotic assortment of mainly brief and individual encounters?

The influence of Electronic Warfare would be in this latter direction, albeit unevenly. The same goes for weaknesses in the infrastructure for operational command and control. These might stem from sheer paucity of resources; and they might also relate to inapposite doctrine and institutions. However, they may be much aggravated by battle damage. Not so long ago we were warned that many of NATO's search-and-control radars were decidedly immobile, and often even lacked anti-blast walls.[66]

Contemporary Swedish practice in this regard affords a useful pointer. The backbone of the semi-automatic STRIL 60 air defence environment is provided by 16 primary radar sets, all of them basically modelled on the US-designed AN/TPS-32. One claim made for this model (peak pulse: 2.2 megawatts on c. 3.0 gigahertz) is that it stands a 90 per cent chance of detecting at 200 nautical miles a perfect target one meter square, set normally across the incoming radar waves; and that it stands a 50 per cent chance of doing this at 300 nautical miles. But since the installations in question must cover a Swedish defence perimeter some 2,000 miles in extent, little room is left for failure in time of conflict. Therefore, the radars are set in rock; and their antennae made retractable. Also, some command centres have been made quite mobile, along with work on decentralising the whole management of air defence. A back-up network of visual observers exists; and there are also contingency plans to incorporate into the system the civilian arrangements for air traffic control.

Yet even a matrix so elaborated is manifestly vulnerable to a sophisticated attack, were this to be on any scale. Furthermore, it cannot provide complete coverage at low altitude nor even approach this ideal except at fearful cost. Nor could these realities be circumvented simply by investing in Over-the-Horizon (OTH) radar or in Airborne Early Warning (AEW) aircraft.

How ill-adapted OTH would be to the air-land tactical environment is indicated in Appendix A. With AEW, however, the balance of the argument is less lopsided. Skyraiders adapted to this role were operational with the US and British navies by 1948 and 1952 respectively. Now AEW platforms are widely in use overland. Moreover, the functions of this genre have extended from spotting to tracking and course prediction; and thence to the direction of friendly forces and the provision of ECM cover. Thus far, the chief expressions of this broadening have been the Tupolev-126 that entered service with the Soviet air force around 1968, and the Boeing E-3A operational in the West since 1976. Each of these Airborne Warning and Control Systems (AWACS) is an adaptation of a commercial transport, the Tupolev-114 and the Boeing 707 respectively. The former is now being replaced by a variant of the Ilyushin-76 transport. All three AWACS weigh between 80 and 100 tons.

Nobody denies that AWACS could sometimes contribute a lot to battle management by tracking numerous targets and directing friendly forces (see Chapter 12). Even so, the large and unwieldy machines of this first generation must be inherently vulnerable, even allowing them a substantial capacity for ECM self-protection. Operating forward, maybe in support of interdiction strikes, they would surely be at risk whenever a conflict was intensive. Located to rearward, they would be put at risk once the battle lines of air combat became broken, irregular or fluid or if their flanks were suddenly turned. Nor dare one discount the dependence of such machines on specialised and limited base support (see Chapter 13).

As always, the course of events in actual war would be shaped by a whole variety of contingent circumstances: the width and depth of the war zone; the mass and density of the air power committed; the electronic balance; the weather, and so on. All the same, it is hard to avoid the impression overall that a major air encounter could readily become a mêlée. There are clear implications for the philosophy of fighter design. But perhaps, too, one should draw the more radical doctrinal inference that the control of air assets should be devolved much more than has customarily been thought right and proper. If this be a valid thought, it is one that complements the arguments that can be adduced (see Chapter 13) in favour of engaging low-flying intruders from positions not so very far above them.

References

1. Field Manual 100-5, *Operations* (Department of the Army, Washington, 1976), ch. 8.
2. John Bartholomew, *The Edinburgh World Atlas* (John Bartholomew, Edinburgh, 1963), p. 17.

160 The Classic Missions

3. FM 100-5, *Operations* (Department of the Army, Washington, 1976), ch. 14.
4. Ibid., pp. 13-14.
5. M. R. D. Foot, *Observer*, 11 March 1979.
6. W. Seth Carus, AIPAC Paper 12, *The Threat to Israel's Air Bases* (American Israel Public Affairs Committee, Washington, 1985), ch. 4.
7. R. A. Mason, *RUSI Journal*, vol. 123, no. 4, December (1978), p. 71.
8. Enrico Po, 'Self-Propelled AA Gun Systems', *Military Technology*, vol. VI, no. 2, February (1982), pp. 29-40.
9. *International Defense Review*, vol. 10, no. 4, August (1977), p. 742.
10. J. M. Collins, *Report to the Senate Armed Services' Committee* (Presidio Press, San Rafael, 1978), p. 281.
11. G. Lowe, 'The Gepard AA Tank', *International Defense Review*, no. 6 (1977), pp. 1083-6.
12. *The Economist*, 4 May 1985.
13. Ronald T. Pretty (ed.), *Jane's Weapons Systems, 1983-4* (Jane's Publishing Co., London, 1983), pp. 91-2.
14. Brian Wanstall, 'Stand-off to Strike and Survive', *Interavia*, 7 (1984), pp. 673-4.
15. Dr J. E. Henderson and Sir Denis Smallwood, *Air Pictorial*, vol. 42, no. 12, December (1980), p. 468.
16. Brigadier General Clifton F. van Kann, 'Vulnerability of Army Aircraft', *Military Review*, vol. XLI, no. 11, November (1961), pp. 2-8.
17. *Flight International*, vol. 113, no. 3608, 13 May 1978, p. 1445.
18. *Aviation Week and Space Technology*, vol. 106, no. 25, 20 June 1977, p. 91.
19. *Para Bellum*, Year 1, no. 12, December (1977), p. 6.
20. David Harvey, 'Dangerous Midnight', *Defense and Foreign Affairs*, vol. VIII, no. 12, November (1980), pp. 12-43.
21. Mauro Zorgno, 'The Laser for Tactical Military Operations', *Military Technology*, no. 6, June (1985), pp. 194-9.
22. For parallels with the sphere of strategic defence, see the paper presented by the author during the 6th National Conference on Quantum Electronics held at the University of Sussex in September 1983. As is also recorded in Chapter 11, it was reproduced in *Navy International*, vol. 88, no. 11, November (1983), pp. 676-81.
23. Benjamin F. Schemmer, 'North Korea Buries Its Aircraft, Guns, Submarines, and Radars Inside Granite', *Armed Forces Journal International*, vol. 121, no. 13, August (1984), pp. 94-7.
24. J. C. Slessor, *Air Power and Armies* (Oxford University Press, London, 1936), p. 144.
25. Solly Zuckerman, *From Apes to Warlords* (Hamish Hamilton, London, 1978), p. 301.
26. R. van Pohl quoted in Eugene Emme (ed.), *The Impact of Air Power* (Van Nostrand, Princeton, 1959), pp. 250-60.
27. T. N. Dupuy, *Numbers, Predictions and War*, (Macdonald and Jane's, London, 1979), p. 94.
28. Purnell's, *History of the Second World War*, vol. 3, no. 10, p. 1165.
29. James T. Stewart, *Air Power: The Decisive Force in Korea* (Van Nostrand, Princeton, 1957), Preface.
30. David Rees, *Korea, the Limited War* (Macmillan, London, 1964), p. 378.
31. Raphael Littauer and Norman Uphoff (eds), *The Air War in Indo-China* (Beacon Press, Boston, 1972), p. 71.
32. Ibid., p. 73.
33. 'Modernisation of Luftwaffe to Counter Soviet Shift', *Aviation Week and Space Technology*, vol. 103, no. 9, 3 March 1975, p. 13.
34. *Field Manual 101-10* (US Department of the Army, Washington DC, 1959), Part 1, Section 7:15.

35. 'The Armour Busters', *Flight International*, vol. 3750, no. 119, 21 March 1981, pp. 815–17.
36. Wing-Commander Duncan Allison, 'Air Support of the Land Battle in the Late 1970s', *RUSI Journal*, vol. 117, no. 666, June (1972), pp. 21–9.
37. *International Defense Review*, vol. 14, no. 6 (1981), p. 765.
38. Donald J. Alberts, 'An Alternative View of Air Interdiction', *Air University Review*, vol. 32, no. 5, July–August (1981), pp. 31–44.
39. Roy Wagner (ed.) and Leland Fetzer (trans.), *The Soviet Air Force in World War II* (David & Charles, Newton Abbot, 1974), ch. 19.
40. John Chapman, 'New Eyes for NATO', *Military Technology*, vol. VII, Issue 10 (1983), pp. 42–5.
41. 'Compass Cope Seen in Anti-Tank Role', *Aviation Week and Space Technology*, vol. 26, no. 21, 23 May 1977, p. 15.
42. Arthur Reed, *Unmanned Aircraft* (Brassey's, London, 1979), pp. 79–80.
43. Paul Dickson, *The Electronic Battlefield* (Marion Boyars, London, 1977), p. 188.
44. *Outer Space — Battlefield of the Future?* (SIPRI, Stockholm, 1978), p. 26.
45. 'Space Sensor Planned to Detect Cruise Missiles', *Aviation Week and Space Technology*, vol. 120, no. 13, 28 March 1984, p. 16.
46. 'USAF Spurs Spaceplane Research', *Aviation Week and Space Technology*, vol. 120, no. 13, 28 March 1984, pp. 16–18.
47. Based on a personal visit, in 1981, to the European Centre for Medium-Range Weather Forecasting at Shinfield Park in Britain.
48. FM 100–5, *Operations* (Department of the Army, Washington DC, 1976), ch. 13.
49. See Frank Barnaby, 'How Death Became Invisible' [sic], *Guardian*, 26 April 1984.
50. Fred Majdalany, *The Battle of Alamein* (Weidenfeld and Nicolson, London, 1965), pp. 73–4.
51. *Soviet Military Power* (US Government Printing Office, Washington DC, 1981), p. 37.
52. See, e.g., Ernst Noack, 'Drones and RPVs Today and Tomorrow', *Military Technology*, vol. VIII, Issue 10 (1984), pp. 14–27, and Michael C. Dunn, 'Bringing 'Em Back Alive', *Defense and Foreign Affairs*, vol. XII, no. 5, May (1984), pp. 24–7.
53. FM 100–5, *Operations*, Section 7–13.
54. Ken Nölde, 'Tactical Strike Reconnaissance: Airpower's Oldest Profession', *Defense and Foreign Affairs*, vol. VIII, no. 10, October (1980), pp. 14–38.
55. *Insight on the Middle East War* (André Deutsch, London, 1975), p. 110.
56. Arnd Brauch, 'Camouflage: The Whole Spectrum', *Military Technology*, vol. VIII, Issue 4 (1984), pp. 24–30.
57. Graham Rose, 'The Design of Future Cockpits for High Performance Fighter Aircraft', *Aeronautical Journal*, vol. 82, no. 808, April (1978), pp. 159–65.
58. Mike Hirst, 'The Next Generation Fighters', *International Defense Review*, vol 17, no. 8 (1984), pp. 1039–44.
59. *Military Technology*, vol. VIII, Issue 5 (1984), p. 41.
60. Arthur Tedder, *The 1947 Lees Knowles Lectures* (Hodder and Stoughton, London, 1948), p. 35.
61. Alvin Perrin, 'In-Flight Refuelling. A Key to Operational Effectiveness', *International Defense Review*, vol. 11, no. 4 (1978), pp. 565–8.
62. *International Defense Review*, vol. 10, no. 5 (1977), p. 860.
63. In the technical literature, the pulse repetition rate is normally entitled the Pulse Repetition Frequency (PRF). This is unobjectionable, of course, so long as it is understood that this usage of 'frequency' has nothing to do with the signification of a point on the electromagnetic spectrum.

64. *Aviation Week and Space Technology*, vol. 114, no. 11, 16 March 1981, pp. 53–6.

65. E. H. Sims, *Fighter Tactics and Strategy, 1914–1970* (Cassell, London, 1972), p. 179.

66. 'Ground Environment for the Future: the Air Defence of Allied Command Europe', *Royal Air Force Quarterly*, vol. 18, no. 4, Winter (1978), pp. 345–52.

8 TACTICAL AIR MOBILITY

One thing that can be said with utter confidence about modern air defence is this. It stands to be at its most effective against monoplane transports: slow, cumbersome and lacking in ECM as they habitually are. Nevertheless many armies still retain specialist airborne troops at battalion strength or above, their hallmark being an ability to assault from such planes by parachute. Among the armies thus provided are all the ones in the Warsaw Pact plus a clear majority of those within NATO. Among them, too, are those of Algeria, Argentina, Bolivia, Brazil, Colombia, the People's Republic of Congo, Dominica, Ecuador, Egypt, Ethiopia, Guatemala, India, Indonesia, Iran, Israel, Japan, Jordan, both North and South Korea, Mexico, Morocco, Nepal, Saudi Arabia, South Africa, Spain, the Sudan, Syria, Thailand, Tunisia and Yugoslavia.

In addition, there are not a few countries that, while they may not boast airborne battalions or brigades, do retain in their order of battle at least one formation of commandoes. As a general rule, too, any such unit will be well adapted to mobility by air. Also they will possess at least some capability for parachute assault, usually in the form of earmarked companies or squadrons. Meanwhile, individuals who have qualified as parachutists may be found throughout an army.

One respect in which airborne units commend themselves to governments relates only indirectly to an ability to drop from the sky. It is that paratroops make very dependable soldiers in law and order crises. They will most likely have volunteered, and then been rigorously selected and trained, for a singularly exacting primary role. They must expect to fight disadvantaged in terms of numbers, heavy weapons, local mobility, logistic support, local familiarity and tactical isolation. On account of what this connotes apropos of being zealous yet adaptive, governments may be disposed to see such men as their ultimate guarantors, against threats from within as well as without.

Take the repeated deployment of paratroops to contain US domestic tension. Little Rock, 1957; Oxford, 1962; Selma, 1963; and Detroit, 1967 are well-known examples. Let us consider, too, the policy of the Chinese People's Republic — a country which does not, after all, envisage the aerial projection of military force beyond its own borders. There are 130 line divisions in the 'main forces' of the People's Liberation Army (PLA). Yet only twelve are held in strategic reserve in central China, in the military region based on Wuhan in the Yangtse valley. But alongside them stand, apparently under air force control, all three of the airborne divisions established in the 1950s. One could add that, during the creation of this airborne echelon, much was made of its including a women's parachute regiment. In other words, it was accorded a special place in the 'learn from the PLA' propaganda theme.

Likewise, airborne troops are often given especially tough tasks to perform in overground frontal attack and — even more so — frontal defence. Thus Israeli paratroops led the capture of east Jerusalem in 1967. Likewise, British paratroops made a key contribution to regaining the Falklands in 1982. When the Germans broke into the Ardennes in December 1944, Eisenhower moved up the 18th Airborne Corps from Rheims; and the subsequent defence by its 101st Airborne division of the nodal though beleaguered town of Bastogne has become celebrated. Similarly, as the Wehrmacht fell onto the defensive almost everywhere, it looked to its paratroops where obstinate resistance was needed. Thanks to the haste with which — from the autumn of 1943 — no fewer than seven extra paratroop divisions were built up from cadres, the results could sometimes disappoint. On the other hand, such episodes as the defence of Breslau and of Brest were heroic even if forlorn.

So airborne troops might still be seen as valuable even if it was thought they would never again use their parachutes in anger. Nor should one rule out the active employment in deterrence or defence of this special facility. The 'strategic insertion' of paratroops may buttress local situations other troops could not reach as quickly if at all. For instance, part of Wehrmacht's 1st Parachute Division was compactly dropped immediately ahead of the British 50th Division as the latter advanced on Catania in Sicily in July 1943. Today, the first phase in an emergency deployment in support of a friendly regime somewhere in Afro-Asia might well be a paradrop to consolidate control of a key airfield.

The rather cautionary lessons from World War II about the offensive use of airborne forces were adumbrated in Chapter 1. They need hardly be qualified much in the light of experience in the immediate post-war period. In October 1950, and again in March 1951, the United States army in Korea conducted paradrops of regimental strength to try and cut off North Korean troops falling back in divisional strength. In each case, only a limited success was registered. But during the covertly co-ordinated campaign against Egypt late in 1956 by Britain and France on the one hand and Israel on the other, paratroops were effective enough against an army that had lost its own air cover early on and which may have been overawed by the prestige and assurance of the states arraigned against it. Successful British and French paradrops took place around Port Said the day before the main amphibious landings. Meanwhile, the chief parachute attack mounted by the Israelis, the dropping of just half a battalion just east of the Mitla Pass, neatly covered their main armoured advance across the Sinai. Come the 1967 war, however, the helicopter almost entirely displaced the parachute as a means of tactical air mobility. Already in Indo-China, in fact, this transition was almost complete. Between 1946 and 1954, French Union troops had dropped in some 150 separate operations. But in the next great conflict there, United States recourse to it was to be only occasional and almost always on a small scale. As Moshe Dayan remarked in 1966, 'American warfare in Vietnam is primarily helicopter warfare.'[1] Nor throughout the war did the South Vietnamese do more than 30 unit paradrops.

The timing and extent of this switch of mode owed a lot to the 1950s being the decade in which the concept of the helicopter at last found mature expression, half a century after the first free flight of such a machine. The speed record for helicopters rose from 123 km/h in 1937 to 200 km/h in 1948, then 356 km/h by 1970.[2] Meanwhile, the Korean war had stimulated awareness of their tactical potential, never mind that by far the biggest role assumed by the 400 used in that campaign was the evacuation of casualties to Mobile Army Surgical Hospital (MASH) units. Several years later, the French in Algeria were to be pacemakers in the application by helicopters of suppressive firepower and, indeed, in the launching of heliborne assaults. Duly, at Fort Rucker in Alabama in 1956–7, the US Army began testing the tactical concept of 'sky-cavalry' equipped with armed helicopters.

Even if no rotorcraft of any kind had ever been invented, however, the *modus operandi* of airborne forces would have been up for fundamental reconsideration in the 1960s and beyond. All else apart, the technique of parachuting has evolved too little since the parachute first came into regular use in 1916. Even today, men cannot jump at all safely except from aircraft flying at an altitude of at least 400 ft and at a speed not in excess of 125 mph, and with an ambient wind speed of 35 mph at the most.

Optimally, the assault transports of World War II could drop men about 20 yards apart. With better exit doors and so on, this distance may nowadays be halved. Yet even this implies a 'stick' of troops of company strength, say, extending a good half mile. It also implies, if company it be, a dropping time of something like a minute. Throughout this interval the parent aircraft will have to be flown very straight, steady and slow at what could be termed a 'dead man's' altitude. Evidently, too, a mere second's delay, through nervousness or whatever, may increase the actual gap between two individuals to over 50 yards. Then again, even a relatively high wind or air speed or altitude may lead to undue unit dispersal.

On the other hand, well-trained individuals jumping from, say, 3,000 ft may paraglide several miles to land almost exactly on a chosen point. Another ruse is to jump from, say, 15,000 ft and open the 'chute at perhaps 3,000 ft. These stratagems do not allow of a coherent formation descent but may give individuals a good chance of entering hostile territory unobserved.

Unfortunately, the difficulty of well-grouped arrival is aggravated by the need for separate delivery of certain supplies. My own recollection is of being on board a Beverley medium-transport the RAF was using to drop provisions onto a narrow airstrip hacked out of primary jungle by British troops screening the contested border between Sarawak and Indonesia early in 1966. With not a hint of enemy interference and in ideal weather, some six runs were made over the target, every one of them quite low and very slow. Yet out of a dozen loads dropped, only four landed directly on that airstrip.

Moreover, there is a signal respect in which the assault delivery of men and — more particularly — materials has receded since 1945. It is that the assault glider is no more. Some 50 different examples of this genre of aerial vehicle were developed between 1935 and 1950. Altogether, over 20,000 such craft were built. The largest type, the Messerschmitt-321, had a 165 ft wing-span and could carry either 200 troops or else rather fewer men plus, say, a medium tank. Likewise, a British-built big glider, the Hamilcar, could carry a light tank, while even so standard a model as the Horsa (also of British design) could lift 6,000 lb of equipment or supplies. Always, however, both a glider and its tug were pathetically vulnerable to good air defences. Besides which, a glider needed several hundred yards of tolerably level ground on which to land. Therefore, it was often easy for the defence to ruin prospective sites through the erection of wooden poles or whatever. Accordingly, with the withdrawal from service in the Soviet air force (in 1955 or thereabouts) of the Yakolev-14, the era of assault gliding closed.

All the same, the delivery of a tolerable scale and diversity of equipment and supplies has remained imperative. One response has been to foster the development of Short Take-Off and Landing (STOL) powered monoplane transports. Another has been to carry further the construction of stressed platforms. Some used these days descend beneath as many as 16 large parachutes, thereby bearing fairly gently earthwards single cargoes of up to 60,000 lb. With the heavier loads, retrorockets may be used to ease the impact on landing.

Yet another remedy — notably in the United States, Canada and Britain — has been Ultra-Low Level Aerial Delivery. With this, platforms or bundles are pulled through a rear hatch by a parachute while the plane moves slowly across the dropping zone at a height of perhaps a few metres. This method can deliver largish loads a lot more accurately than can the other modes of heavy drop. However, the training costs are high. Besides, it is hard to blend with a more classic backdrop and is perilous under fire.

Account can be taken of these various solutions. So can it of the way in which some bulky items — among them medium pieces of artillery — may be dismantled for dispatch, then assembled on arrival. Also airborne forces stand to benefit from the trend towards military equipment becoming more compact in relation to performance. Among the most relevant examples are radios and radar, minelets and certain kinds of anti-tank or anti-aircraft weapons. Likewise, the consumption of ammunition, as measured by weight, is falling in relation to its effects.

The fact remains, however, that many individual loads are hard or impossible to drop by parachute or even to deliver by the aircraft actually landing. After all, just a landrover and trailer together weigh 8,000 lb. Then again, the M-551 Sheridan light tank, in service with the US Army, weighs 35,000 lb, while the battle-ready weight of the new US Main Battle Tank, the M-1 Abrams, is 125,000 lb. The M-60 the latter is displacing weighs 11,000 lb. Its Soviet counterpart, the T-62, weighs 90,000 lb.

So today it is still difficult to transport solely by air an adequately full and balanced spread of military impedimenta, let alone paradrop all of it at the further end. The massive Lockheed C-5 Starlifter strategic transport can just manage two M-60s, while the Antonov-22 can carry a single T-62. The C-130 Hercules, which can be rated either a strategic or an intra-theatre transport, can lift a couple of Scorpions (British light tanks, weighing but 17,000 lb) or a 155 mm towed howitzer. Correspondingly, these larger air transports can variously move quite a comprehensive assortment of other kinds of equipment: tracked SAM launchers, air traffic control facilities, powerful and versatile wireless transceivers, armoured recovery vehicles, and so on. All the same, commanders would properly be reluctant to send C-5s, let us say, over or even close to enemy lines. Besides, there are certain items — such as heavy bridging equipment and tank transporters — that seem to be ruled out permanently.

The actual payloads borne on a given occasion will greatly depend on fuel requirements. Floor strength may also be a limiting factor with armoured vehicles etc. Otherwise, the shape and size of the aircraft' hold may be. 'Soft-skinned' vehicles usually exhibit, even when loaded, a density of 10–12 lb for every cubic foot they encompass. With armour, the figure only rises to between 20 lb and 30 lb, still under half the density of water. Admittedly, the figures arrived at by dividing the respective 'cargo envelopes' in aircraft holds by their maximum payloads have typically been in the 5–10 lb bracket.[3] What has to be remembered, however, is that the shapes of vehicles and other cargo are usually such as to leave a good deal of empty space around them.

This is conspicuously true of, for example, a stowed helicopter, even when its rotor blades can be either folded or detached. Witness the awkwardness with which even a machine as small as a Lynx (see Appendix C) fits into a Hercules. In some cases, however, auxiliary fuel tanks may be fitted into cargo compartments or else on flanges in order to make a helicopter independently mobile over a stage length of a thousand miles or more. Also it is becoming easier to adapt rotorcraft to being refuelled in flight. In this context, however, fuel consumption may not be as critical a limiting factor as crew fatigue.

Tactical Air Mobility 169

Everything considered, any strategic airlift must be expected to move a lot less, over a given distance and within a given time, than might be inferred from a naive reckoning of aircraft numbers, payloads, ranges and speeds. A calculation was done by the Pentagon in 1977 on the assumption of the total commitment of its Military Airlift and, of course, of sufficient suitable airfields at the receiving end. The result was 160,000 tonnes for a 30-day inter-theatre movement; and of this total, 28 per cent comprised outsize items eligible for movement only in the C-5s. Also, the official advice then was that a single US armoured division (plus basic supplies) weighed 50,000 tonnes complete. A mechanised division weighed 45,000 tonnes and an infantry 26,500 tonnes.[4] Yet even these figures exclude any provision for the support echelons behind the division proper. Often these are as big, manpower-wise at least, as the division itself.

Apart from them, however, the major formations just cited would be a lot (perhaps two or three times) heavier than their counterparts in World War II. Here then is a factor that may aggravate much further the strain that air transport fleets are subject to. A salutary comparison is that between the Hercules of today and the C-47 Dakota of World War II fame. The maximum payload of the former is four times what the latter's was. On the other hand, only 1,700 Hercules have entered military service world-wide, whereas over 10,000 Dakotas did.

Clearly, then, the constraints that may operate against assault by parachute are closely paralleled by those against the use of air transport to make armies strategically mobile. In theatres like Germany, however, weaknesses in this latter respect can be offset by (a) the pre-stocking of equipment and stores; and (b) a rising reliance on sealift as crisis reinforcement proceeds. So the most urgent question remains how far the role the parachutist assumed in World War II can be taken over or even extended by the heliborne infanteer.

Heavy lift may be an appropriate point of departure. Already some of the rotorcraft in service or prospect can carry quite large individual loads. Confirmation of the entry into Soviet air force service of the Mil Mi-26 (now the world's largest helicopter) came in 1983. Also, moves have been afoot to resurrect, at least as a technology demonstrator, Boeing Vertol's 65 tonne Heavy Helicopter (HLH). The latter could lift an underslung unit load of 35 tonnes while the Mi-26 can manage one of 20 tonnes. By comparison, let us consider a model designed for amphibious assault. Sikorsky's CH-53E Super Stallion weighs 33 tonnes all-up, of which the actual payload may be 14.5.[5] Alternatively, up to 55 troops may be packed in. Sometimes, too, a pair of rotorcraft lift a cargo platform.

As regards tactical movement on an extensive scale, however, a major constraint is the logistic support the helicopters themselves demand. In the course of a ten-hour day a single CH-47A Chinook, say, could deliver about 25 tons of mixed stores across a distance of 200 miles. But in so doing, it would consume nine tons of aviation spirit and occupy, more or less directly, some 50 men. A Caribou fixed-wing tactical transport would deliver 12 tons, consume 3.5 tons of fuel and employ roughly the same number of men. However, five 5-ton trucks travelling at 25 mph on good, straight and direct roads could do best of all. Between them they could move a similar weight of stores, yet need 1.5 tons of fuel and generate work equivalent to that of just 15 men full-time.[6]

Still, nobody claims that the case for the helicopter can be made via cost-benefit analysis as trite as that. The real point is that there will be occasions, even in theatres as well-developed as is central Europe, when neither a motor-truck nor even an aeroplane as agile as the Caribou can arrive sufficiently close to where men or supplies are urgently destined for. Furthermore, a local road network that otherwise would adequately sustain a certain military commitment might be vitiated by the destruction of bridges or culverts plus the presence of wrecked vehicles, columns of refugees or, for that matter, snow and ice. In the light of World War II and Korean experience, US Army planners worked out the typical capacities, *vis à vis* the transit of useful loads, of two-lane roads. One with a bituminous surface and set in rolling terrain could, in a combat situation, move up to 5,000 tons per day, enough to sustain several divisions fighting quite hard. But seasonal bad weather alone could be expected to cut this norm by 30 per cent. And if the routeway in question were but earthen, the representative daily rate would drop from 1,200 to 120 tons.[7] Bulk movement across rough ground may, of course, be even more weather dependent.

Movement over any kind of land surface is also susceptible to scattered minelets, and to other forms of air and artillery interdiction. But transit by helicopter may avoid these several threats, largely or completely. Over and beyond which, an ability to travel between two points some ten times as fast as any land vehicle comes strongly into its own in heading off enemy breakthroughs. A

Tactical Air Mobility 171

further advantage of the rotorcraft, in any combat situation, is that it can also be used for surveillance or as a weapons platform.

Take, for example, the AH-64 Apache (see Appendix C). This is designed most specifically for low-profile attack, especially against armour and especially through the use of the 16 Hellfire missiles it can carry on a single sortie. The two-man crew:

> can employ Hellfire in one of three basic ways: (1) autonomously, with the co-pilot/gunner designating the target with the laser, (2) indirectly, in coordination with a ground-based designator, (3) in cooperation with an airborne laser designator in a scout helicopter or another Apache. These modes can be used at night or in bad weather, as well as during the day. The Apache can, if necessary, remain hidden by terrain cover, launching its Hellfires unguided against targets up to approximately ten kilometres away, to be designated *en route* by another laser.[8]

Suffice to add that, in due course, external target designation might be carried out via RPVs.

All of which serves to emphasise how much scope attack helicopters have acquired for firing from shielded and even 'hull-down' positions. They can do so with assurance by virtue of sights and designators set either on masts on their own structure or else located separately. It is an attribute which may do much to shape the future of the air–land battle (see Chapter 13). Already, in fact, various analyses have indicated that, in these circumstances, (a) a high proportion (perhaps 85 or 90 per cent) of the missiles directed against enemy tanks by the rotorcraft will hit their targets; (b) the tanks' ability to return fire will be very limited; and (c) no other weapons systems will readily be brought to bear on the helicopters from the ground.[9] Still, all concerned do realise that such craft can be threatened from the air either by other rotorcraft or by monoplanes. This is not least because their own rotors are highly visible, when revolving, to doppler radar. What may also merit more consideration is how far the scenario just outlined depends on the assumption that, as a general rule, a Forward Edge of the Battle Area (FEBA) can be clearly identified. Once a conflict has degenerated into a mêlée (perhaps in part because of 'vertical envelopments' by assault rotorcraft), the immunity being sought through launching flank attacks from hull-down positions over friendly territory may elude the helicopter crews.

Soviet helicopter losses in such circumstances in Afghanistan have been high but hard to collate. So one turns first to South Vietnam to observe what may be the prospects in a chaotic war. By early 1971, over 4,000 helicopters had been lost by US and allied forces, either in accidents or through adversary action. Still, that denoted a loss rate of, in the US case, only one per 7,000 missions. However, that ratio is to be seen in relation to (a) many of the flights in question having been over areas of low military activity; and (b) a lack within the Communist ranks of any surface-to-air weapons more potent than 12.7 mm heavy machine guns. Nor should one forget that no fewer than 15,000 helicopters had been recovered from operational areas, after sustaining either accidents or breakdowns or battle-damage.[10]

Agreed, it can fairly be argued that, while the hilly jungles and wet paddy of Vietnam made helicopters indispensable, they also left them peculiarly exposed both to incidental mishaps and to small-arms fire. Yet exposure to fire was even more evident in October 1973, in the bare and flat environs of the Suez Canal. None the less, the Egyptians say they cast no fewer than 50 commando detachments across the Sinai in the first afternoon of that campaign. True, this force projection was effected partly by traditional parachute drops. But it was effected largely by two or three dozen helicopters, each bearing some 30 men. Broadly, the aim was to generate confusion.

However, a good third of these craft were eliminated forthwith. In fact, the Israelis claim that 'Thirty-five Egyptian helicopters, many of them loaded with troops, were destroyed in the first days of the war'.[11] The then commander of the Israeli air force was later to dismiss such machines as 'clutch' weapons with 'no right to exist' on a modern battlefield.

But Germany and, to an extent, Korea have blends of topography, vegetation and building that are more conducive to evasion through 'Nap-of-the-Earth' (NOE) tactical flying. On the other hand, a FEBA in Germany, in particular, might quickly become fluid and fragmentary. So it is necessary to consider a helicopter's prospects, as and when it is exposed to direct fire from cannon or something worse.

One way to survive can be to win a fire fight. Helicopters can concentrate rapidly, and so take full advantage of the square law. What is more, an individual machine may, depending on the model, be capable of fierce suppressive fire. Thus the 30 mm chain gun an Apache bears fires 625 rounds a minute. Also it can carry eight anti-tank guided missiles or up to 76 unguided 2.75 in rockets. The latter can be fired in a brief time span, their combined impact being equivalent to a salvo from a World War II light cruiser. Then again, helicopters may carry air-to-air missiles; radar-warning receivers; radar or infra-red jammers; and chaff or flare dispensers.

Another factor in favour of safe return may be image compression, at any rate in head-on profile. Thus the Gazelle anti-tank helicopter operated by the French army has a frontal silhouette of but 2 sq.m, which makes it hard even to see at a few kilometres. Contrariwise, noise and shudder (especially when at the hover) could make so tight a vehicle a poor platform for observation over that sort of distance. But at least a Gazelle can be equipped with a stabilised observation sight.

Agility as an aid to evasion is best measured in terms of initial acceleration along the horizontal and vertical axes. As a rule, however, top speeds are more readily quotable. The Apache has a maximum rate of climb of 762 m a minute, though this can be cut by half by even a modest warload or, to think more grandly, dispatch to tropical uplands. What is more, no helicopter is able to pull much more than 3.0 g. This means that, at a speed of 290 km/h, say, it could not turn a semi-circle with a diameter much under 400 m. Still, that distance compares well enough with those characteristic of warplanes flying tactically.

As with warplanes, one of the biggest problems is fragility. After all, 60 kg of steel armour or 45 kg of laminate may be needed to protect a square metre of surface against bursts of fire from even a light machine-gun.[12] Still, it is claimed for the Apache that both airframe and main rotor are tolerably resistant to impacts by single 23 mm cannon shells. Also, the two crew members are shielded against such ordnance by seat armour plus a vertical barrier, interposed between them. But then one has to remember, too, how readily automatic weapons of heavier calibre might be trained on such a machine. Besides, the chief threat to crew safety is the loss through battle-damage of aerodynamic lift. Seat armour or no, they would be lucky to survive a plummet to Earth of more than 20 m.

Adverse weather also constricts the use of helicopters, though perhaps less than it does that of monoplanes. Icing of the rotor blades and ice ingestion by the engines yield only slowly to intensive research, hence the general proscription on flying rotorcraft in snow and ice that still obtains in the USA.[13] Nor is this surprising, seeing that rotor-blade 'wing-loadings' (wing area divided into all-up weight) are always extremely high and that rotary motion compounds the resultant stress.

On the other hand, the extra difficulties posed by tactical flying at night, be it foul or clear, are of diminishing consequence. In 1972, a full test of 'Nap-of-the-Earth' (NOE) flying in darkness was conducted at the Hunter Liggett Military Reservation in California, using in-service helicopters and with central Europe's landscapes particularly in mind. With something like a full moon, NOE could continue at close to the daylight norm — that is to say, not more than 50 ft above the treetops or whatever. But with a new moon, NOE had to be conducted at up to 125 ft, a relative altitude that was much more perilous. Moreover, free manoeuvring would often be infeasible below 250 ft. Added to which were big problems with target discrimination.[14]

These last few years, however, the introduction of superior equipment (e.g., radar altimeters, Forward-Looking Infra-Red scanners, and image-discerning goggles) has made rotorcraft less susceptible to poor light. Not long ago, only an eighth of the helicopter exercises in Western Europe were taking place between 22.00 and 05.00.[15] Soon, it should everywhere be possible to operate between these times on the great majority of occasions.[16]

Yet whatever the ambient circumstances, crewing a helicopter remains a most stressful responsibility, even by aeronautical standards. Naturally, this is largely because of the intricacy and subtlety of the actual task. However, it is also due to vibration and noise. But just as task performance has been aided more directly by science, so, too, has this incremental stress been diminishing. Witness the reduction of the mean vibration within their rotorcraft claimed by Sikorsky. It has fallen from 0.70 g in 1945 to 0.39 g in 1965 and 0.10 g in 1975. From which there are also dividends in terms of the efficiency, longevity and reliability of component parts.

Ignoring for now the roles of the helicopter at sea, the sense one is left with is of the future of rotorcraft within the land battle hinging on several delicately-balanced assessments. Some of these relate to characteristics inherent in rotorcraft as a genre. Others may be subtly dependent on the evolving 'state of the art' in adjacent areas of technology. Some relate to alternative V/STOL modes, including those hybrid as between pure helicopters and the classical fixed-wing form. Others may concern the tactical use of the parachute. Then again, others derive from the evolution of land warfare as a whole and, in particular, of low-level air defence.

Undeniably, the helicopter can be much at risk from surface anti-aircraft fire. Alternatively, however, it may complement it. At present, this contribution is mainly seen as snap interceptions or dogfights against other helicopters.[17] But low-level defence against fixed-wing aircraft may be on the horizon. One avant garde but authoritative view has been as follows: 'With proper weapons and employment, it is an outstanding tactical air-to-air weapon against fighter and attack aircraft, forced by a surface-to-air threat to operate in airspace most advantageous to helicopter capabilities.'[18]

With all these dimensions to the debate, a more synoptic consideration is best deferred to Chapter 13. Enough for the moment to look a little further at how thinking on the overland employment of military helicopters has progressed within the Superpowers, the two states that already own five or ten thousand apiece. In the USA, the way through was the evolution of the 'air mobility' concept. A leap forward came in 1962, when Secretary of Defense McNamara established a senior board (chaired by Major-General H. H. Howse, the first Director of US Army Aviation) to explore 'new organisational and operational concepts, these perhaps including completely air-mobile infantry, artillery, anti-tank and reconnaissance units'. The ensuing 'Howse Report' advocated more use of rotorcraft throughout the army and called for the formation of five 'air-assault divisions', each of which would possess 459 helicopters together with a number of monoplanes. Otherwise, the said formations were to be rather lightly equipped. So as occasion demanded, they were to employ some of their helicopters as 'gunships' in order to supplement the Honest John rockets and 105 mm howitzers of their divisional artillery. Moreover, both those weapons could be made 'heliportable'.[19]

The first such division was inaugurated at Fort Benning early in 1963. Even so, only two Airmobile Divisions were ever deployed by the US Army in Vietnam: the 1st Cavalry and the 101st Airborne. The former was activated in 1965 and the latter converted in 1968. Their respective titles well epitomose the fusion thus of two exotic traditions.

Now the 9,000 helicopters (plus 525 monoplanes) in the US Army are spread broadly across the service at large.[20] There is one air assault division and one independent air cavalry brigade. The other 15 formed divisions each possess *c.* 175 helicopters, some three-quarters of which are organic to a combat aviation battalion. Just over a third of its machines are of the light observation kind, the remainder being apportioned almost equally between attack and transport/general-purpose varieties.

In traditional Soviet practice, helicopters are part of the air force. But the inventory, now risen to over 4,000, is mainly organised into special regiments, one of which is attached to every first-line army. Indeed, the Soviet army seems now to be taking over this rotorcraft fleet.

This move is, of course, part and parcel of the complex resorting of Soviet air assets lately in progress. In itself, however, it can fairly be held to betoken a resurgent awareness by Moscow of the battlefield potentialities of such rotorcraft, a resurgence to be related to such notions as the Operational Manoeuvre Group and the primacy of surprise attack (see Chapter 2). Meanwhile, more direct evidence is accumulating that, although the vulnerability of the helicopter does cause concern in high places, Soviet doctrine lays ever more stress on the value of this vehicle as an instrument of assault by vertical envelopment, a form of mobile artillery, a platform for surveillance or command and control, and a minelayer or minesweeper. Procurement levels are authoritatively expected to rise sharply.[21] The advent in 1985 of impressive new designs has enhanced the credibility of this prediction.[22]

References

1. Maurice Tugwell, *Airborne to Battle* (William Kimber, London, 1971), p. 322.
2. John Fay, *The Helicoper* (David and Charles, London, 1976), ch. IX.
3. Neville Brown, *Strategic Mobility* (Chatto and Windus, London, 1963), p. 156.
4. J. M. Collins and A. H. Cordesman, *Imbalance of Power* (Presidio Press, San Rafael, 1978), p. 199.
5. Flight International, vol. 125, no. 3919, 23 June 1984, p. 1619.
6. Brown, *Strategic Mobility*, p. 194.
7. *Field Manual 101-10*, (US Department of the Army, Washington DC, 1959), Part 1, Section 7:15.
8. *Jane's Defence Review*, vol. 4, no. 3 (1983), p. 206.
9. For an eloquently optimistic prevision of the contribution of the helicopter to a defence of West Germany see Colonels J. N. W. Moss and J. L. Waddy, *RUSI Journal*, vol. 129, no. 2, June (1984), pp. 27-32.
10. *The Role of the Helicopter in the Land Battle* (RUSI, London, 1971), p. 38.
11. Chaim Herzog, *The War of Atonement* (Weidenfeld and Nicolson, London, 1975), p. 258.

12. *International Defense Review*, vol. 10, no. 5, October (1977), p. 865.
13. *Flight International*, vol. 119, no. 3753, 11 April 1981, p. 1031.
14. W. L. Odneal, 'Effect of Nap-Of-The-Earth Requirement on Aircrew During Night Helicopter Operations', *Journal of the American Helicopter Society*, vol. 20, no. 4, October 1975, pp. 43–52.
15. *Interavia*, no. 7, July (1980), p. 620.
16. Brian Wanstall, 'Among the Weeds at Night', *Interavia*, no. 7, July (1985), pp. 793–5.
17. Mark Lambert, 'Helicopters Prepare for Air Combat', *Interavia*, 5 (1985), pp. 445–6.
18. Dan Hartley, 'Helicopter Air Combat', *Flight International*, vol. 126, no. 3919, 11 August 1984, pp. 25–7.
19. Lt. Gen. John T. Tolson, *Airmobility, 1961–71* (Department of the Army, Washington DC, 1973), ch. 1.
20. *Military Balance, 1985–86* (The International Institute for Strategic Studies, London, 1983), p 8.
21. C. N. Donnelly, 'The Soviet Helicopter on the Battlefield', *International Defense Review*, vol. 17, no. 5 (1984), pp. 559–66.
22. 'First Havoc, Now Hokum — What role for the new Soviet Helicopter?', *Interavia*, no. 7, July (1980), pp. 798–9.

9 THE VULNERABILITY OF AIRFIELDS

Perhaps the most vexing question of all, in the ramified debate about the future of warplanes, is their susceptibility to immobilisation on the ground. Though bad weather alone may still effect this, the big anxiety these days is pre-emptive enemy action. Either the machines themselves will be wrecked or else their runways or other facilities will.

Yet pressing though this matter seems in relation to modern technology, it has been a source of worry for a long time. When nearly 60 years ago J. F. C. Fuller sardonically yearned for a 'portable landing ground', he characterised the manned aircraft as:

> a kind of animated rocket fired . . . from an aerodrome, a weapon . . . as immobile as a fortress . . . In brief, the future of air power is not to be sought in the air but on the ground, for the ground is the bed and the belly of the monoplane.[1]

Before many years had passed, the critical dependence thus depicted was to be savagely exploited, through air strikes and landings, in successive Nazi blitzkriegs. In Poland, airfield attack was not conclusive, as is explained below. But it was the prime means whereby, within the first three days of the invasion of the Low Countries in May 1940, a useful measure of air superiority was made an overwhelming one.

Several years later, during the Ardennes offensive, a desperate attempt was made to repeat this *coup de main*. On New Year's Day, 1945, 1,000 Luftwaffe warplanes were mustered to try and prevent the Allies taking advantage of a sudden clearance of the fog and low cloud that till then had curtailed their air operations. As dawn broke, this fleet attacked 27 air bases in Belgium and Holland. It thereby destroyed 300 Allied aircraft for the loss of 93 of its own.

In the first hours of the invasion of Russia, 70 Soviet airfields had been bombed. German sources put the number of aircraft wrecked that day at nearly 1,500. Soviet historians concede the loss of 800, explaining how essential improvements to 200 airfields 'located in the new border zone' had only begun a few weeks previously.² But by 1945, the Red Air Force itself was using the tactic to effect: 'It usually took five sorties to destroy an enemy aircraft on the ground . . . as against 30 in air combat'.³

Since then, the most notable examples of hitting thus at the foundations of an 'air umbrella' have come from the Middle East. Within three days of the start of the Suez war in 1956, high-level strikes by RAF bombers had grounded Egypt's air force. Within three hours in 1967, the synchronized strike by Israel against 16 Arab fields was to determine the course of a 'Six-Day War'. Come 1973, however, airfields were lower in her attack priorities. Urgent calls for Close Air Support on the Golan and for interdiction of the canal crossings were the initial reasons. Later on, her adversaries suffered such loss rates in the air, of pilots hardly less than planes, as to make this the preferred way to set them back another decade. Besides which, the Egyptian air force, in particular, had acquired since 1956 much more in the way of base protection, both passive and active. None the less, the Syrian sortie rate was eventually reduced by airfield attack.

Unfortunately, air staffs throughout the world have never squared up to this threat in a sufficiently firm and steady manner. Harsh lessons from World War II about the exposure of planes parked in neat lines had virtually been forgotten by 1967. Then the Arab débacle did revive, more or less everywhere, awareness of the need to shield aircraft in revetments. Soon, too, certain NATO air arms were exercising themselves about the susceptibility of their stations to chemical attack. So this further stimulated revetment construction, along with other improvements in passive protection.

Yet these responses were nowhere comprehensive, except perhaps in North Korea.⁴ Nor was the vulnerability of runways and other infrastructure adequately addressed. Through 1980, indeed, the whole subject seemed to recede. Thus in September 1982, a most successful multinational conference on 'The Future of Air Power' was to be held at Birmingham, under the auspices of the Royal Air Force and our university. The agenda was rich and varied. But the viability of bases was not addressed.

Since then, however, there has been an upwelling of concern — most noticeably in the Pentagon — about reducing dependence on runways. So how sustained and constructive will this trend be? Air arms retain their historic disposition (see Chapter 3) to perceive established air stations not as the exposed nubs of a intricate and rather fragile panoply of air power but as rock-solid citadels.

180 The Vulnerability of Airfields

What is more, some aspects of the record seem prima facie to justify their sometimes being regarded not merely as citadels but as sanctuaries. After all, the UN Command in Korea never attacked the fields in Manchuria from which the MiG-15s came on their interceptor missions. The Falklands was fought and won with Argentina's mainland bases apparently off-limits. Other examples could be mentioned. Correspondingly, the 1976 exposition of NATO air doctrine warned that 'Before conducting or escalating an offensive counter-air campaign, careful consideration should be given to the response' that this might provoke.[5] In other words, such a widening of a war is liable itself to have wider consequences. Airfields situated the far side of cardinal political boundaries may, of course, pose particular problems.

Yet any immunity thereby gained for air bases will always be contingent, never absolute. During the Korean war, pressure insistently built up on the American Right in favour of hitting those Manchurian fields, pressure the White House might have bowed to had there not been tacit *quid pro quos* (e.g., Communist abstention from attacks on UN carrier task forces) or had an acceptable armistice not been signed in 1953 (see Chapter 1). Likewise in 1982, London never overtly ruled out the bombing of airfields within Argentina. Instead, the two strikes by a single Vulcan from distant Ascension Island against the runway at Port Stanley served as an oblique warning that, though sorties against the mainland might be even harder to execute, almost nothing was impossible. Certainly an impression built up in Washington that such attacks might be mounted, an impression that reportedly caused rifts within the National Security Council.[6] Besides, there have since been unofficial intimations that the Royal Marines and/or Special Air Service did quietly operate against the said targets. So maybe fears on this score at that time caused Argentina's air force (a) to disperse in depth more than it otherwise would have, or (b) to hold a proportion of its warplanes back for airfield defence.

Therefore, the historical record cannot be said to lend support to any notion of absolute sanctuary. Often theatre commanders and political leaders will feel inhibited about extending hostilities across national boundaries or otherwise deepening a conflict zone. As and when such inhibitions are overcome, however, this is very likely to be to hit at air bases.

Before proceeding to an examination of the threat aircraft and missiles can thus pose, it is well to mention a related menace that

The Vulnerability of Airfields 181

has likewise been neglected. This is the susceptibility of airfields or airstrips to capture or harassment by advancing enemy land forces. As a rule, neither the location nor the configuration of such a site makes it suitable as a node of local defence. Nor is it only the flow of the main battle that is to be apprehended on this account. World War II experience confirms that airfields also lie peculiarly exposed to attack by roving columns (e.g., Britain's Long-Range Desert Group) or, of course, airborne troops.

This question is one to which the British defence community ought to have much to contribute. Experience in the use of RAF personnel for airfield defence or associated combat tasks on the ground stretches back to the creation of the RAF Regiment in 1942; and, indeed, to the formation, in the Middle East over 60 years ago, of the first RAF armoured-car units. Yet today the RAF Regiment cannot normally provide more than a single light armoured squadron (with six light tanks plus 15 armoured personnel carriers) for the ground defence of a forward airfield. The only other RAF contribution in this sphere would be a static infantry one provided from the airfield itself. At any given time, this would comprise one of the three 'watches' into which all the personnel had been divided for 'round-the-clock' operations in war. No doubt this provision could prevent infiltration by indigenous Leftist fanatics or by squads of Spetznaz, the Soviet special purpose paratroops. But it might prove much too weak in firepower against an assault by one of the USSR's new Operational Manoeuvre Groups (OMG). Such a move might be made to secure an 'airhead', the better to sustain a progressive disruption of the whole rearward area.

All the same, the current debate largely revolves around the susceptibility of air stations to attack by aircraft or missiles. About which, there is this to be said straight away. Fuel and ordnance can well be stored underground, with little to fear except the fragility of the access points. But otherwise, a lot of the facilities on a permanent air base are difficult to harden or replicate in any very adequate way. Among them are hangars of the customary size and structure. So, too, are master radars, not least because they are prone to attract homing missiles. Nor may camouflage do a lot for such installations, except perhaps against missiles under televisual guidance from nearby aircraft. Still, the sheltering of individual aircraft (plus spare parts, etc.) within concrete revetments can afford them excellent protection against a wide diversity of non-nuclear ordnance. Furthermore, when these structures are built as fully enclosed shelters, their sealed doors and associated filters may provide ground and air crewmen near immunity from fallout or, indeed, most lethal chemicals. By late 1982, some 60 per cent of all the USAF planes that, after a week of crisis reinforcement, would be assembled in central Europe could thus be accommodated.[7]

With the progressive introduction of revetments there and elsewhere, the emphasis on runaway spoilation has become all the more pronounced. Witness the purpose-made ordnance reviewed in Chapter 4. Nor can anyone deny that modern runways are inviting targets. They are often a good 9,000 feet long and several hundred feet wide. They have a base layer of rubble, the thickness of which varies (depending on the subsoil) between one foot and several yards. Then there is the top surface of concrete, usually rather less than 2 ft thick. A major problem is that, if a warhead does go clean through and then explodes beneath, it exposes the concrete to deformation through stretching, a form of stress it does not easily withstand.

However, warplanes overflying wide and flat airfields at heights of maybe 20 metres do face a growing risk of being shot down. One factor operative here is the little room they have for manoeuvre *vis-à-vis* the bearing on which they approach. If the angle chosen is much less than 30 degrees from the lie of its axis, the runway may not be crossed at all. If it is greater, the bombs may straddle yet miss. Some measure of the ease with which defending forces may exploit these geometrical constraints is provided by the Rampart passive area-defence lately developed in Britain. Among its components is the Skysnare: a helium-filled kite balloon that is 4.8 m across and which can ascend to its maximum height of 300 m within two minutes. Tests indicate that its tethering cable could all but sever a warplane's wing.

In the face of such challenges, the standard format for airfield attack has to evolve towards penetration by means of missiles that have either been released by aircraft in 'stand-off' positions or else have been fired surface-to-surface or even (e.g., adapted Polaris or Trident C4s)[8] from submerged submarines. A possible variant of the Pershing 2 would carry up to 76 scatterable warheads. Each would weigh 18 lb, a fifth of this being high explosive. Studies have suggested that the dispatch of 100 such missiles might cut the Warsaw Pact sortie rate by as much as 35 per cent.[9] Never mind that this particular ordnance could never penetrate two feet of solid concrete; nor that the US Army (the operator of Pershing 2) is not as yet keen to count runway-busting among its tasks; nor that the Warsaw Pact is bound, in due course, to acquire closely comparable weaponry, given the Soviet penchant for surface-to-surface missiles whether ballistic or 'cruise'.

Clearly, surface-launched missiles have great potential in the counter-airfield context. Even so, they are susceptible to antidotes. For instance, Rampart also includes decoy rockets designed to foil incoming missiles. It includes, too, generators which may cause dense smoke to bloom rapidly, thereby obscuring the ground below for minutes on end. Then again, camouflage may blur the sharp imagery that homing sensors depend on. Furthermore, the close-in protection of airfields may eventually lend itself to laser-beam defence. Even so, the likelihood is that, through the 1990s at least, missilry will be directed with high confidence against air base facilities in general and runways in particular.

Evidently then, there is a pressing need to improve the robustness of airfields as a whole. Presumptively, this must entail, first and foremost, keeping runways open. That, in its turn, chiefly means strengthening their surfaces. Recently, concrete reinforced with fibres (either steel or plastic) has been laid at certain US airports and military air stations; and may exhibit over twice the normal flexure strength.[10] What is more, there appears yet to be considerable scope for strengthening cement as such by making its texture more consistently good. Obviously, too, the existing surfaces could be thickened throughout, at a cost of, say, ten million dollars a runway. Besides, not even a weapon as advanced as the Durandal (see Chapter 4) would stand more than an even chance of piercing completely with a single hit the runway surfaces typically in use today.

Moreover, the technology of rapid runway repair is already quite advanced, as witness a system Britain brought into service in 1972. It involves an aluminium mat being bolted across a crater that has been filled with loose stone. Granted, this can never be as secure a repair as a plug of concrete would be. On the other hand, a mat takes two hours to lay whereas plugs have usually taken several to set. Reportedly, however, the Japanese Defense Agency has evolved a new fluid fill (based on ammonium and polyurethane) that will set to operational requirements within 30 minutes.[11]

Still, the story by no means ends here. A succession of hits may have a cumulative effect that piecemeal repairs cannot obviate, even if resources are to hand. Moreover, the heave deformations caused when dibber bombs do explode under the concrete may be too acute to ignore, yet too broad for quick correction. Likewise, pitting by cluster weapons can slow take-off runs dangerously, as can extended cracks and debris. Finally, the needful length and width of runways render them all too liable to repeated neutralisation by scatterable minelets.

In short, runway use seems doomed to remain too susceptible to disruption and reduction, not to say curtailment. Hence the mounting concern to shorten or even abolish the take-off and landing runs aircraft actually make. In the ultimate, the most radical response must be that whereby machines take off with little or no run, entirely under their own power; and, in due course, descend to a spot touch-down. Of the two recognised paths to this result, one — the helicopter — has evolved in relation to a whole variety of tactical considerations, as per Chapter 8. But the other — the Vertical or Short Take-Off and Landing (V/STOL) monoplane — is promoted with liberation from runways being the overriding aim.

Suffice for the moment to remark how conducive technical and operational trends currently are to more reliance being placed on V/STOL in tactical warplanes. Maximum air speeds are now often at levels which mean that engine thrust will, in any case, be fairly close to a plane's all-up weight. The constraint on fuel load imposed by the special need to limit weight for V/STOL is less acute than before since fuel use is becoming more efficient. Arguably, too, deep strike ought henceforth to command less priority (see Chapters 5 and 13). Also germane are the considerations which militate in favour of more attention being paid to a close-in mode for the 'look-down, shoot-down' interception of low flying intruders (see Chapters 7 and 13).

But VTOL or V/STOL apart, there has been no lack of innovatory response to the anxieties engendered by runway dependence. Some variants of the MiG-21 have long been fitted with a Jet-Assisted Take-Off (JATO) unit — that is, a pair of auxiliary turbines that will be jettisoned just after departure. Similarly by early in the 1960s, the USAF was sometimes fitting a JATO twin on its A-4D Skyhawk. Meanwhile, a few of its F-100 Super Sabres were experimentally lifted off ramps by rockets with thrusts equivalent to 4 g at normal loads. This 'zero-length' launch was a technique with which the British firm of Fairey had experimented, with quite positive results, for several years from 1949. It was also to be tried out on a few of the F-104G Starfighters which the Luftwaffe started to procure from 1963.

By that time, however, tactical warplanes were developing in such a way that their take-off distances were falling well below their natural landing runs from touch-down. Therefore, the latter had become the critical factor. Accordingly, the US Navy and Marine Corps had come to use arrester wires ashore, as well as at sea. It is a remedy which may now be on the verge of a significant extension. Thus the USAF has lately conducted tests of two mobile arrester gears: the BAK-12 and BAK-13. Each can stop successive aircraft, within a few hundred feet, every two or three minutes; and, with the BAK-13 at any rate, a machine thus arrested may weigh as much as 20 tons and have touched down at no less than 185 mph. In the BAK-12, the braking resistance is produced by frictional mechanics. In the BAK-13, it comes from turbines acting against a water-glycol mix.[12]

Moreover, both reverse thrust and parachute braking have long since been normal fits on the front-line planes of many countries; and each can reduce landing runs by 1,000 to 2,000 ft.[13] Already, too, variable-geometry wings (able to extend to maximise lift for either take-off or landing) are integral to certain important designs. Correspondingly, the last two decades have seen a general reduction in the speeds on touch-down of high-performance warplanes. Whereas the landing speed of both the F-104 and the Mirage III (each of which entered service 25 years ago) is 225 mph, that of the F-16 as well as the Tornado is 135 mph. This attribute (plus powerful engines to take off with) allows the Tornado to operate off less than 3,000 ft of runway despite an all-up weight of 24 tons.

All the same, concern has often been expressed over the singular facility with which Warsaw Pact air arms have operated off treated or even rough grass. In the case of the MiG-21, JATO has certainly played its part. Otherwise, low weights, wide tyres, and articulated undercarriages have ensured this dexterity. However, the edge the Pact has in this regard diminishes as heavier warplanes enter service with it. The loaded weight of a MiG-25 is over three times that of any MiG-21, while that of a Sukhoi-19 is over twice a Su-7's.

None the less, aviation circles in the West remain as anxious as ever about one particular corrective. This is limiting the pressure exerted per unit area through the tyres of their aircraft. Four years ago, in fact, the USAF signed a development contract for something called FloTrack: a flexible surface to be clipped round an aircraft's wheel so as to lessen by perhaps two-thirds the pressure exerted on soft or rough ground during taxiing.[14] In addition, the next generation of USAF fighters will be fitted with much tougher landing gear to enable them to cope better with damaged surfaces.[15]

What does have to be cautioned against, however, is too strong a disposition to rely in war on whatever odd stretches of a standard long runway happen to be temporarily and relatively damage-free. These may well lie, extremely exposed to weather and to recurrent enemy harassment, somewhere out in the middle of airfields which may also have suffered attrition of other facilities, radar and wireless very possibly included. The least to expect in consequence is a lowering of sortie rates and a decline in tactical co-ordination. Surely that alone is argument enough for bursting more decisively the bonds of the air-base syndrome. In other words, exploit some of the technical possibilities just discussed to disperse squadrons, early in a warlike emergency, to sites that (a) cater for fewer aircraft apiece, and (b) are not heavily structured and permanent.

How helpful such a distribution can be may be seen from Poland's air strategy in 1939. As one officer has put it:

> In retrospect, it seems quite naive of the Germans to have believed that, during the preceding days of high political tension . . . we would leave our units sitting in their peacetime locations. . . . In the previous 48 hours, all of us had transferred to emergency airstrips.[16]

The upshot was that, even after 17 days of bitter and ever less equal conflict, over a quarter of the 450 aircraft which the Poles had started with could be flown out to Romania.

No doubt this remedy was made easier by many of the machines in question being light and of low performance, even by the lights of the time. But it was facilitated, too, by the widespread Polish woodlands. Likewise since 1959 the Swedes have turned their distinctive northern geography to positive advantage by having many of their front-line aircraft mainly operate from designated stretches of highway, mostly widened or strengthened for the purpose. On landing, 'Aircraft are dispersed to purpose-built hardstandings or on suitable side-roads in woods.'[17] The proud claim is that this pattern enables Royal Swedish Air Force (RSAF) fighters to conduct 'between two and three sorties in the time it takes other air forces to carry out one and prepare aircraft for the next'.[18] The

The Vulnerability of Airfields   187

Finns, too, apply this precept. Indeed, it stood them in excellent stead during the Russo–Finnish War of 1939/40.[19] Likewise, in 1973 the Syrians flew some combat missions from a prepared highway strip near Damascus in 1973, as their airfields came under Israeli pressure.

Also NATO has identified within the Federal Republic over 100 stretches of autobahn and the like that could sustain Harriers, Jaguars, A-10 Thunderbolt 2s and so on. In addition, the USAF has been preparing Forward Operating Locations (FOL), each for wartime use by eight A-10s normally based in Britain. Every FOL will have several concrete shelters and a wartime complement of 100 men.

Usually, dispersion to civil aerodromes is a further alternative. In Britain, for example, there are 65 with hard runways at least 2,000 m in length. True, civil and military activities are never entirely congruent. In particular, a lack of active and passive defence together with other discordances (apropos of radio frequencies, quality of fuel, etc.) have tended to make civil airfields far from ideal in actual war. Nevertheless, they have figured in sundry contingency plans for tactical aviation as well as for air transports and strategic bombers.

Lastly, one argument in favour of reverting to light and less sophisticated warplanes is that these may be able to operate off short and rough grass strips. During the Falklands war, between 20 and 50 locations on the islands were usable by the 30 Pucara light strike-aircraft the Argentinians possessed. Instructive also is the way the British built a 250 m aluminium strip near San Carlos to enable Harriers to refuel and then stand by.

Some of the objections to dispersal have either been exaggerated all along or else become weaker over time. Clearly, the shift in the balance of operational advantage towards VTOL (see above) is an aspect of the weakening. Then again, the improving reliability of components (not least electronic ones) should gradually ease the logistic strains engendered by trying to service warplanes spread in 'penny packets' in often awkward locations. Nor should we forget that a lack of hardened facilities and of local defence may be offset by the use not just of camouflage but of mock-ups (see Chapter 7). What these stratagems do not preclude, however, is the assured pinpointing of sites by the tracking of aircraft as they leave or return.

What is more, the 'command and control' problem is aggravated by dispersion, not least because the electronic environment may be flipping abruptly and constantly between coherence and chaos. Originally, the Luftwaffe drew up a plan to implement extensively its concept (mentioned above) of launching Starfighters from ramps. However, it was cancelled in the light of a USAF study which had concluded, in 1963, that the strain of dispersion on the command structure could be excessive.[20] It is a factor that must especially weigh with nuclear or chemical release.

Virtually inherent in the very notion of small and well dispersed sites is the tenet that bad weather will be less tolerable yet often more prevalent than on a regular airfield. Besides which, automatic landing and other aids to all-weather operation will best be afforded on established airfields. Needless to say, the same applies to the handling of those vital yet dangerous consumables: fuel and ordnance. Evidence of what holdings of these are deemed necessary is not easy to come by. But we know, for example, that a modern French air base near Cambrai is building up a week's supply of ammunition.[21] We have also been told that each of the air bases the Saudi air force has been creating will have enough fuel for 30 days on the basis of a modest two sorties per day.[22] At a dispersed or improvised site, the norm could be set a lot lower. Even so, it could hardly be below a day's worth. Yet a squadron of twelve Harriers completing, say, five sorties in a 24-hour day could well go through 500 tons of fuel and ordnance.

Above all, however, one has to insist that a degree of ground dispersion cannot confer immunity just like that. It certainly cannot against modern surveillance. As much is recognised by the RSAF with its new 'Base 90' dispersal plan, field tests of which began in 1985. The aim is to curb what could otherwise be a rising risk of immobilisation on the ground, even with dispersion. One hazard identified is that the infra-red emissions from warm engine-casings, etc. could be more detectable as sensors improve. Another is simply that the number of front-line warplanes in RSAF service is due to fall sharply over the next decade or so. Between about 1992 and 2005, virtually all the squadrons in question will either re-equip with the JAS 39 Gripen (now under development at Saab) or else they will disband. The Gripen inventory may never exceed 240, whereas 425 front-line warplanes are in service today.

The authors of 'Base 90' envisage mobile maintenance units (comprising nine or ten men apiece), one or more of which would operate from each of the several satellite baselets that would be earmarked for use in crisis by the squadrons on a permanent air station. Within every baselet area would be some 50 hardstands able to accommodate a single aircraft apiece. These individual dispersal points would regularly be used for servicing and take-off. Some might even be suitable for landings. Otherwise, an aircraft would land on a satellite runway (which would be narrow and not over 3,000 ft long) and then taxi across.[23]

However, the imperative of keeping the Gripen easy to maintain under circumstances such as these may be constricting its overall development, as witness a reported decision to dispense with a thrust reverser. Worse, the sheer elaboration the Base 90 way of doing things involves would surely (a) increase manpower needs; (b) increase the risk of discovery through the interception of wireless traffic; and (c) ruin the impressive rate of sortie generation the RSAF has sustained to date.

Nor, in the light of so much historical experience, dare anyone ever ignore the possibility of an air arm being caught unprepared on its peacetime bases. This circumstance could arise in three ways. One is attack straight out of the blue, with the prevention of dispersion as perhaps a main aim. The second is that, although warlike preparations by the adversary have been observed, it has been felt that reciprocation would be alarmist or provocative. The third is that regional tensions have stayed so high for so long that continued dispersion has become unacceptable for military and civilians alike.

Yet acknowledgement of the difficulties that surround dispersion as a stratagem should not lead one back to complacent acceptance of the status quo. Vulnerability is getting worse. It affects not only the planes which will fight the theatre air-land battle but also those concerned with strategic reactions and with maritime support. It calls for radical solutions.

References

1. J. F. C. Fuller, *On Future Warfare* (Sifton Praed, London, 1928), p. 202.
2. Roy Wagner (ed.) and Leland Fetzer (trans.), *The Soviet Air Force in World War Two* (David and Charles, Newton Abbot, 1974), p. 19.
3. Y. Yeraksa, *Soviet Military Review*, 10, October (1975), p. 49.
4. Benjamin F. Schemmer, 'North Korea Buries Its Aircraft, Guns, Submarines, and Radars Inside Granite', *Armed Forces Journal International*, vol. 3, no. 8, August (1984), pp. 94–7.

5. ATP-33, *NATO Tactical Air Doctrine* (NATO, Brussels, 1976), p. 408.
6. Jeremy Campbell, *The Standard*, 18 May 1982.
7. Donald E. Emerson, P-6810, *USAFE Airbase Operations in a Wartime Environment* (Rand Corporation, Santa Monica, 1982), p. 8.
8. *Report by the Military Committee's sub-committee on Defence Co-operation on the Implications of Technology for the Battlefield* (The International Secretariat of the North Atlantic Assembly, Brussels, 1982), p. 7.
9. *Flight International*, vol. 120, no. 3770, 8 August 1981, p. 434.
10. M. Vandevalle, 'The Fast Evolution of Steel Fiber Concrete in Airport Applications', *Aerospace International*, vol. 6, no. 2, March (1980), pp. 120–3.
11. *Aviation Week and Space Technology*, vol. 118, no. 19, 9 May 1983, p. 11.
12. 'USAF Tests Arresting Systems', *Aviation Week and Space Technology*, vol. 120, no. 1, 2 January 1982, p. 45.
13. See the author's *Strategic Mobility* (Chatto and Windus for Institute for Strategic Studies, London, 1963), ch. VII, pt 3.
14. *Flight International*, vol. 121, no. 3808, 1 May 1982, p. 1180.
15. *Flight International*, vol. 124, no. 3885, 22 October 1983, p. 1114.
16. Quoted in Charles Messenger, *The Art of the Blitzkrieg* (Ian Allan, London, 1976), p. 130.
17. *Flight International*, vol. 123, no. 3858, 16 April 1983, p. 1042.
18. *International Defense Review*, vol. 16, no. 4 (1983), p. 419.
19. William Green, 'Finland's Modest Air Arm', *Flying Review International*, vol. 24, no. 4 (1968), pp. 48–53.
20. J. S. Tompkins, *The Weapons of World War Three* (Doubleday, New York, 1964), pp. 77–8.
21. *International Defense Review*, 5, 1985, p. 576.
22. *Aviation Week and Space Technology*, vol. 118, no. 21, 23 May 1983, p. 45.
23. 'Sweden Adjusts to Military Reductions', *Aviation Week and Space Technology*, vol. 120, no. 4, 20 January 1984, pp. 101–12.

PART THREE

OTHER ENVIRONMENTS

10 AIR POWER AT SEA

A moral may be drawn from the fact that several of the early successes of operational research were on the interface between air power and naval (see Chapter 1). This is that these two domains of war can readily be bracketed. In each, offensive action is less constrained by the physical environment than it usually is on land. Weaker armies deploying for defence have, from time immemorial, used the landscape to buttress themselves. In air or sea battles, on the other hand, the attack has usually found it easier to exploit an initial edge, be this in terms of mass or of quality. The constraints imposed by geography are less acute.

By much the same token, both aerial and maritime conflicts still appear to lend themselves to analysis and prediction more readily than does warfare on land. Above all, each seems more amenable to numerical assessment than ground fighting is ever thought to be. Hence today, navies no less than air forces look for big dividends, proscriptive and educational, from computerised war games.[1] Another element in this correspondence is that the control exercised by technology over tactics and through them strategy looms large in both cases.

Still, this eligibility for discussion in broadly similar terms does not mean that much the same conclusions may be drawn about the impact of the newest technologies on manned aircraft, be these landbased or shipborne, and on naval vessels. Let us consider, for example, the sophistication of modern marine mines. It is true that every aircraft that operates off a ship is thereby at some risk from these weapons. Nevertheless, it is safe to say that the marked progress the marine mine has lately made has much improved the status of the aircraft. It has done so relative not only to the surface ship but also to the submarine. After all, mines pose a considerable threat to the latter especially when laid in the channels leading to their home ports. Moreover, aircraft are well suited to offensive minelaying in inshore or, of course, inland waters. Then again, though monoplanes have never adapted at all well to minesweeping of any kind, helicopters may regularly be employed to trigger certain kinds of influence mines as well as to cut the cables that anchor contact ones. Witness the encouraging experiments the US Navy has lately done on having a helicopter tow through the water a magnetic-mine sweep.[2] Still, unmanned surface vessels or submarines, remotely controlled by some parent craft, are also in prospect as an alternative minesweeping mode.

The intricacies of mine warfare apart, however, both the submarine and the aircraft are making headway as against the warship. Accordingly, they now vie with one another as the chief determinants of future maritime supremacy. Turning then to the submarine, the mix of factors which affect its status in a limited war at sea cannot be quite the same as what will be adumbrated in Chapter 11 apropos strategic nuclear deterrence. Sometimes more critical in the more localised scenarios will be the gains in submerged speed, depth of travel and agility afforded by nuclear propulsion and the novel hull-designs associated with it. However, no nuclear-driven submarine yet in service displaces less than 2,400 tons. Nor is this lower limit likely to come down very fast or far until such time, perhaps around the year 2020, as unmanned versions enter service. Until then, all such boats will be (a) expensive and (b) difficult to use in shallow waters.

When attacking ships from at or near the surface, submarines will henceforward place less reliance on their classic weapon — the heavy torpedo — this in favour of missiles which mainly travel through the atmosphere, often over long distances. In the face of this threat alone, the contribution that aircraft make to Anti-Submarine Warfare (ASW) must be central to any assessment of their future in the maritime environment as a whole.

Once again, however, neat summation is impossible. Geography in all its aspects is always a determinant. Besides which, there is an acknowledged need regularly to employ in ASW a mix of systems (monoplanes, helicopters, surface ships and submarines). Therefore, the whole question becomes one of striking appropriate balances rather than making absolute choices. After all, ASW has to be conducted around the often turbulent interface between two quite different media, the atmosphere and the sea. Moreover, certain properties of the latter (salinity, temperature, oxidation) vary widely and often sharply from place to place and time to time, especially in the vertical plane and within a thousand feet of the ocean surface.

These variations inevitably impede what remains, none the less, quite the most useful means of anti-submarine surveillance, the acoustic or sonar. Two principles are alternatively applied. They are the 'active' transmission of sound pulses (so as to bounce them off target bodies) and the 'passive' reception and recognition of noise emanating from the target itself. The inputs of energy demanded by active sonar are substantial. Not least is this because, in the active mode, the weakening of emissions through propagation is in both directions. On the other hand, the time a pulse takes to return can be measured in order to estimate the range to target: something the passive mode can only yield (and then but roughly) as and when two platforms prove able to conduct a triangulation. Furthermore, the pulses from active sonar may be distinctive even against a background of high ambient noise, whereas the emanations from a target will always be diffused in frequency and strength. Still, computerised records of the noise signatures of all the prospective sources has made obscuration easier to cope with. Besides which, passive sonar never betrays the presence and bearing of its parent platform. But an active one may all too easily do just that.

Sonar transducers can be fixed in surface ships, submarines, or moored buoys or be placed on the seabed itself. They may also be dipped beneath the waves from ships or hovering helicopters. In addition, free-floating sonar buoys (passive or active) may be dropped in suitable search patterns from monoplanes. Unfortunately, however, acoustic surveillance by either kind of aerial vehicle is acutely subject to the usual constraints on weight and volume. What is more, the acoustic sensors themselves rest on or near the surface when borne in buoys released from monoplanes, and do not dip far beneath if suspended from helicopters. In other words, neither method avoids the waters near the surface: waters which are invariably more disturbed and through which sound waves may fail to pass from lower levels on account of refraction and reflection. This avoidance is achieved only by (a) Variable Depth Sonar (VDS) sets installed in escort ships displacing several thousand tons apiece, or (b) another submarine operating at appropriate depth. What needs to be remembered is that a nuclear-driven adversary boat travelling at, say, 35 knots can descend 1,500 feet within half a minute.

Not that a hunter-killer submarine patrolling to a similar depth necessarily stands the best chance of making the actual kill. A major problem for it will be communication with the other elements in a task force. In fact, the one way available will be to use its transducers as a primitive transmitter of morse, maybe across a few miles. Another difficulty is that, when such a vessel is deep down, a heavy torpedo is effectively the one weapon it can employ. Yet here a problem of relative speed presents itself. Reportedly, the USA's new Mark 48 heavy torpedo can manage 55 knots. Yet even this improvement impresses only until one recalls that the USSR's new Alfa boats reach 43 knots, over twice the World War II best for a submarine.[3]

Weaponry that flies through the atmosphere for much of its path may therefore be more likely to catch its prey. Still, such weaponry needs to include missiles big enough then to traverse hundreds of feet of water to deliver a powerful explosive charge. Here one is talking about devices that can be borne only on large monoplanes or escort ships.

A serious objection to the submarine as an ASW platform is that it may face counterattack, perhaps from the very boats it is hunting. This at any rate cannot be said of ASW aircraft, except in two respects. Once again, the most obvious is that the smaller kinds may have to work from warships which themselves are vulnerably placed. The other is that the advent of Vertical Launch Systems (VLS) may eventually make it feasible to fit a submarine out with an anti-air ability more menacing than anything tried in such vessels heretofore, a theme lately underscored by USN experiments.

Meanwhile, the most fundamental drawback of the ASW ship derives directly from the physical dynamics of the sea-air interface. The fraction of the energy generated by its engines that is elegantly but uselessly absorbed causing bow waves rises ever more steeply as the vessel's speed through the water mounts. Take the external resistance engendered by a fleet destroyer under way. The proportion from the said wave formation is liable to rise from 30 per cent at 16 knots to 80 per cent at 27 knots. Likewise, the total drag at 24 knots is nearly four times what it is at 16 knots, while at 27 knots it is 20 times.[4] Among the adverse consequences is that even active sonar is progressively degraded at speeds above 15 knots; and, at the present state of the acoustic art, has become almost useless by the time 28 knots has been reached.[5] All of which virtually precludes a destroyer's keeping up with a nuclear submarine.

An added complication is that — in sea areas as crowded as, say, much of the Atlantic always is in peacetime and might still be in war — background noise may be most troublesome. Needless to say, too, repeated explosions could exacerbate matters: 'An entire ocean can be rung like a bell to cause it to reverberate from shore to shore.'⁶ Clearly, acoustic sensoring can thus be impaired at the longer ranges. At the same time, however, the noise emanating from large and fast modern vessels may vitiate the classic defensive ploy of sailing closely in company:

> Bunching of ships . . . tends to eliminate effective responses to close-in submarine datums. If passive acoustic measures are used to detect a submarine, the noise-field created by many ships travelling close together tends to drown out submarine-radiated noise. When active acoustic measures are used, the multitude of near targets and their wakes create a host of false and confusing targets — hindering rapid localisation of a submarine contact at the longer follow-up ranges involved.⁷

Here then is a dialectic between dispersion and concentration that may be resolved only by ASW platforms that may proceed to an assignment far more swiftly, over extended distances, than even a nuclear-driven submarine. An escort ship cannot raise steam enough for this; and the helicopter it will normally bear nowadays often fails to offer enough of an added dimension to compensate. Duly, some analysts have set store by the potential of the hovercraft as an ASW spearhead. Riding on an air cushion the way a hovercraft does is quite heavy on fuel but does permit speeds of 75 knots and more as well as good sonar reception.

What ought to be more urgently considered, however, is whether the advent of the ASW hovercraft should not be paralleled by a more determined development of large Long-Range Maritime Patrol (LRMP) aircraft. Granted, the latter cannot exploit sonar technology as freely as the former can. On the other hand, an LRMP may cost an order of magnitude less to build than would an ASW hovercraft of, say, 2,000 tons. Furthermore, it would be better at covering large surface areas, with its high maximum speed and its use of radar to detect protruding periscopes or, of course, fully surfaced boats. Then again, only aircraft (whether monoplanes or rotorcraft) are suitable as platforms for Magnetic Anomaly Detectors: sensors that can fix the exact location of submarines at distances up to 2,000 ft. To all of which should be added the evident correspondence, conceptually speaking, between the large LRMP geared to ASW and the Long-Range Combat Aircraft mentioned in Chapter 11.

One sphere in which the analytical thrust is towards smaller aircraft again (see Chapter 12) is Airborne Early Warning (AEW). By this is meant the forewarning of aerial intrusion that naval air arms (meaning, in the first instance, the American and British) pioneered after 1945. Followed the Falklands war, navy circles in London and Washington discussed anew how to provide this facility from short flight decks. An AEW version of the Harrier VTOL fighter was contemplated, either a two-seat or else a single-seat that would be confined to data-transmission.[8] Another possibility examined has been the installation in an American tilt-rotor airframe of a radar developed for the RAF's Nimrod four-engined AEW machine.[9] Already, too, helicopters play at least a limited part.[10] Moreover, one approach considered has been to make an in-line pair of rotor blades serve as a radar antenna of quite exceptional horizontal aperture. Maybe the most fundamental reservation to express is that a rate of rotation of around a thousand degrees per second would cause a blurring of echo return that simple calculation suggests would be unacceptable (even for non-doppler registration) beyond a range of 50 miles.

The prospects for effective AEW, against low-flying aircraft or missiles, are enhanced over the open sea by 'ground-clutter' interference being less pronounced there, at most radar frequencies. Witness all the speculation in the West about the 'look-down' capability of the Tupolev-126. There has been broad agreement throughout that this Soviet AWACS is at its best over water.[11]

Without AEW, ships are the more at risk of being overwhelmed by enemy aircraft or missiles closing at supersonic speeds and just above the surface. After all, the vertical build of warships allows the siting of radar antennae only sufficiently far above the waterline to present a geometric horizon some 10 or 20 miles away. Yet an antenna revolving, say, five times a minute will only register a body approaching at, say, 1,200 miles an hour every fourth mile of its flight. So there is little margin for threat assessment. Besides, if an antenna locks onto the one contact, others may go by default.

On the other hand, the ducting of electromagnetic energy brought about by temperature or humidity variations within the lower atmosphere can extend the scan of surface radar, maybe several times over. But ducts are commonest in low latitudes. In such sea areas as the Persian Gulf or the Bay of Bengal, in fact, they may be encountered on a majority of occasions, sometimes to altitudes of several thousand feet.

Assuming enough forewarning, the acquisition radar which is

integral to the Goalkeeper, a modern Dutch naval gun of 30 mm calibre, enables it to destroy supersonic aerial vehicles at distances between 1,500 m and 400 m. Two such aircraft or missiles can be engaged simultaneously, if observed on much the same quarter.[12] Missile launchers can similarly be used for self-defence by ships proceeding singly or in company. However, they cannot 'gather' missiles into their guidance beam in so short a minimum distance.

What should not be forgotten is that, even in World War II, attacks delivered at low altitude and point-blank range against surface ships could veritably be suicidal. So it was that, in the European theatre in the middle of that conflict, the standard parameters for personnel planning within the RAF called for '95 per cent replacement of the aircrew of a torpedo squadron on operations every three months'.[13]

Admittedly the situation eased a lot later, so far as the Allies were concerned. But this was because a steady shift in the numerical balance allowed the regular commitment of planes *en masse* in order to achieve a swamping of a convoy's defensive screen along with the direction of much of the offensive effort to that screen's forceful suppression, mainly with cannon fire. Meanwhile, the progressive displacement of torpedoes by rockets was bestowing some freedom of manoeuvre (as regards speed, direction and altitude) on the planes assigned to the prime strike mission.

Since 1944, however, the quality of anti-aircraft defence has improved as markedly in the maritime sphere as elsewhere. Nor may the numerical balance at sea as a rule be as favourable to aircraft in the future. In any case, a plane closing at very low level with a compact surface target may have to rise to several hundred feet to orient itself for the actual strike. Hence the interest naval air arms have sustained, these last 25 years or so, in 'toss-bombing'. By this is meant utilising the sheer momentum of the attacking aircraft to hurl a bomb a mile or two forward as the machine itself rises into a backward loop: an involution also intended to avoid passing over the close-defence zone in question.

A toss-bomb (probably set to explode some 150 ft above the surface) is unlikely to sink even a Fast Patrol Boat (FPB), not with wind-drift and all the other parameters so well judged as to attain perfect accuracy. What it may do, however, is injure enough of the crew or damage so badly radar scanners, missile launchers and other external fittings as to leave the objective wide open to a follow-up attack in more orthodox style. Therefore, this option does remain of account.[14]

But more central to the debate about maritime air power is the use of the longish-range missile, launched from a submarine or warship or aircraft. An ability to skim along barely above the sea surface is a singular attribute of the cruise missile. What with this, their compactness and the fact that they may be programmed for rather erratic flight, such weapons are not easy to shoot down with the assurance required to protect ships. The best answer may be close-in defence with automatic cannon.

Their value as platforms for the mass-dispatch of surface-to-surface cruise missiles is one of two reasons (a perceived requirement for coastal bombardment being the other) why the USN is recommissioning no less that three Iowa-class battleships. Yet enough has already been said to show that the air-to-surface cruise missile, released from a large aircraft standing a long way off the target area, will also come into its own in maritime warfare. Thus it is mainly for use against shipping that the Soviet family of eight such weapons has been evolved. Western estimates of their operational ranges vary. Still, the upper limit for the AS-3 which is borne on the massive Tupolev 20 (see Appendix C), appears to be 650 km while for the AS-5, carried by the smaller Tupolev 16, it is 320.

Contrariwise, most of those naval SAM systems customarily described as 'longer-range' are never effective beyond slant distances of 30 or 40 km. Even the largest of these weapons, the Talos (which is installed in UN warships of cruiser size) is propelled by its ram-jet motor across a slant range of but 120 km. Part of the explanation lies in the altitude contrast between monoplane and ship. More importantly, however, the former platform is always the more mobile. A supersonic aircraft might move well over 100 km while a Talos was chasing it, a prospect likely to make interception futile.

This is the chief reason why a major task force typically has to extend itself across what may approximate to a box some 300–500 miles square. Only thus can the larger and more precious of its ships be shielded from stand-off attack by deep zones of defensive fire. However, a company of ships spread that wide may find it hard to deploy in the narrower seas, the Eastern Mediterranean for instance. What is more, a force optimised, in both composition and density, for warding off aircraft or missiles may be less well adapted to the exclusion of submarines or fast patrol boats.

Not that it is all that easy to devise a configuration that will be viable against air attack alone. After all, the density of defence is bound to be low compared with what would be regarded as normal — not to say imperative — by many commanders on land. Sheer cost limits the number of vessels. Also there are severe constraints on the amount of anti-aircraft coverage any one of them can muster. All warships also have other roles to fulfil. They do, in any case, devote a high proportion of their substance simply to keeping afloat and mobile.

Consider, in particular, a frigate acting as a radar picket in a forward location. It may be isolated from support but without adequate means of self-protection. Yet on its survival may depend the integrity of the task-force as a whole. Here, after all, is the dilemma the USN found so bitter as, during the battle for Okinawa in 1945, it faced nearly 2,000 attacks by *kamikaze* suicide bombers: 'manned cruise-missiles', as one might say. It was this experience, above all others, that gave rise to AEW.

For the task force as a whole, concealment and deception may confer a measure of immunity. Bad weather can be exploited to effect. So can radar reflectors intended to mislead as to which of the ships in company are big or otherwise important. So can active electronic jamming coupled with physical evasion. Above all, chaff may better screen ships than it does planes. Mainly, this is because the differences between the movement of the chaff and that of the craft it is shielding will be less with the former. On occasion, too, a rough sea may similarly help.

Apparently, during the Falklands campaign, four Argentinian air-launched Exocet missiles were enticed away from their intended objectives by broadcast chaff or by helicopter-borne emitters. However, one of these Exocets then went on to sink a container ship, the *Atlantic Conveyor*. In addition, Exocets were successfully aimed against two destroyers, sinking one and seriously damaging the other. One of the missiles in question had been launched from the air and the other from the land.[15]

So apprehension of the eventual arrival of missiles that can perform even better has contributed to a heightening of interest in reversing the recent trend to bigger warships. Nor is this just a matter of fewer men and less material being lost when a ship goes down. It is also because a smaller vessel is better able to evade attack, not least air attack, by changing speed and course or through deception. Gradually, too, we will see the application, in small designs to start with, of new hydrodynamic hulls.

Meanwhile, manifold technical changes now in train should make the disadvantages of small scale less acute in all sorts of ways. Already seaworthiness is less of a problem at low displacements. Then again, the progressive miniaturisation of electronics counts for a lot at sea. Take, for example, the L-70 gun installed in the Spica FPBs of the Royal Swedish Navy (RSN). A Spica has a standard displacement of only 190 tons. Yet the L-70 is a stabilised and radar-directed weapon able to engage either surface or aerial targets with 57 mm shells, this at a firing rate of 200 rounds a minute. Little wonder then that the RSN sees no need to acquire further warships of destroyer size and above. Likewise, the USN's Harpoon missile, with its range of 60 miles, is eligible for fitting on any combat vessel except a patrol gunboat.

Not that the virtues of largeness have entirely vanished. There are still a number of weapons and support systems that can be carried only in the big ships. What must also be said, certainly with present-day hulls, is that the engine power needed to sustain a given speed falls off sharply in proportion to ship tonnage as that tonnage rises. Let us compare, for instance, two USN vessels each credited with a maximum speed of 33 knots: a Carrier Vessel Attack, the *Forrestal*, and a guided-missile destroyer, the *John Paul Jones*. The former has 19 times the displacement of the latter but its main engines have only to be 3.7 times as powerful. Added to which are the technical economies of scale that make large engines a good deal more efficient, weight for weight.

Regardless of any general balance of advantage, however, there are two categories of warship in which the attractions of size remain more compelling than before. The one is the amphibious assault ship, the reason being that several navies continue the World War II innovation of using big vessels as strategically-mobile 'mother-ships' for assault boats or — these days — helicopters. But the other is the aircraft carrier.

Here a prime consideration has been the sheer size of the aircraft they now have on board. Let us make a representative comparison. The six carriers with which the Combined Japanese Fleet attacked Pearl Harbour displaced between 16,000 and 38,000 tons standard; and yet they operated between 65 and 85 aircraft apiece. The 13 carriers in the United States Navy today range in standard tonnage from 51,000 to 91,000. Even so, their aircraft complements are between 75 and 95, a decidedly marginal change in relation to the contrast in ship sizes.

Not that the era of the supercarrier has been accepted without dissent. The late 1940s saw bitter controversy in Washington about the strategic utility of the *Forrestals*. In the early 1960s, in Washington and soon thereafter in London, the debate about the utility of large fleet carriers resurged, especially in relation to limited war outside NATO. Numerical comparisons were made, notably as between the *Essex* class of 33,000 tons standard displacement and the *Forrestal* of 76,000. The latter seemed to win hands down, even though it boasted only 25 per cent more planes. For one thing, the differential was 50 per cent when account was taken just of those aircraft earmarked for attack missions. Also, for stocks of ordnance it was over 200 per cent. Furthermore, the *Forrestal* launch rate was over twice as fast, and the turn-round time a quarter less. The accident rate was half. Correspondingly, the reckoning was that, in storm-swept waters like the Norwegian Sea or the Straits of Taiwan, a *Forrestal* was truly operational 345 days a year as against 220 with an *Essex*.[16]

All the same, anxiety is always engendered by the heavy concentration of human and material resources even an *Essex* represents. After all, 15 per cent of those *kamikaze* sorties off Okinawa ended with a hit against some vessel or other. Could a big modern carrier justifiably be exposed to a similar onslaught with cruise-missiles or whatever?

In formulating a reply, the first essential is not to be misled by the seeming ease with which carriers were sunk in World War II. Outright sinkings of that kind of ship might not be effected as readily in the future. Nearly half the 25 carriers then lost by the various navies were destroyed by fire or major explosion, a hazard since reduced by the advent of jet aircraft with their less volatile fuel (see Chapter 13).[17] Meanwhile, big advances have been made in the limitation of battle damage, not least by armouring flight decks and installing more watertight bulkheads. An early pointer in this direction was the resilience displayed, in her death throes in 1943, by a battlecruiser, the *Scharnhorst*. She did not actually sink until she had taken hits by innumerable shells and a dozen torpedoes. A modern aircraft carrier might do just as well in that respect.

However, it was the penetration of her forward boiler room by a single shell (of 600 kg) that ruined the *Scharnhorst*'s tactical prospects. So the moral is surely this. The capital ship (then the battleship or battlecruiser; and today the carrier) may be hard to annihilate but all too easy to cripple. What this means in practice will depend on circumstances and, of course, luck. But even one missile homing onto a key radar, say, might turn so magnificent a vessel into more of an obligation than an asset.

Nor dare one ignore the disruptive implications for embarked air squadrons of almost any damage to their ship. Here the rub is that operation off carriers has always been within terribly narrow margins of room and time. Thus even the flight deck of the USN's nuclear-driven *Enterprise* (76,000 tons) looks utterly congested with but half the ship's full complement of planes parked there as economically as possible.[18] Nor should one forget the accidental fire that broke out in the *Forrestal* on active duty off Vietnam, a fire that claimed 129 lives as well as 57 aircraft destroyed or badly damaged. All in all, the carrier remains altogether too liable to sudden transformation from trump card to millstone.

Nor is there all that much scope for sacrificing some of the technical advantages of scale in order to distribute risk (and maybe improve area coverage) by going hard for a greater number of much lighter carriers. The smallest vessel currently operating a squadron of fixed-wing aircraft is the Spanish *Dédalo*. But though the nine planes it bears are Harrier VTOL machines, this ship still displaces 13,000 tons. That makes it a good five times as big as the average destroyer and frigate in the Spanish navy.

At present, a broad consensus among the experts would appear to be as follows. Only the Superpowers can hope ever to operate enough really large carriers to dominate any likely opposition. However, it is going to remain decidedly impracticable to build carriers as such below the displacement bracket of 10–20,000 tons. True, there will be scope for distributing VTOL machines (monoplanes or rotorcraft) throughout a navy and also on merchant ships. Even so, the upshot is that land-based aircraft (and land-launched missiles) are likely to bear more heavily than before on naval events.

References

1. 'Navy Using Computerised War Games', *Aviation Week and Space Technology*, vol. 120, no. 21, 21 May 1984, pp. 181–7.
2. *Military Technology*, vol. VIII, Issue 2 (1984), p. 68.
3. Norman Friedman, 'Modern torpedoes — a new generation?', *Military Technology*, vol. IX, Issue 7 (1985), pp. 28–37.
4. D. Phillips-Birt as quoted in the author's *Strategic Mobility* (Chatto and Windus for IISS, London, 1963), p. 143.
5. R. E. Adler, 'In the Navy's Future: the Small Fast Surface Ship', *United*

States Naval Institute Proceedings, vol. 104, 3, no. 901, March (1978), pp. 102-11.

6. V. C. Anderson in B. T. Feld *et al.*, *The Impact of New Technologies on the Arms Race* (MIT Pugwash Memorandum, Cambridge, 1971), p. 203.

7. W. J. Ruhe, 'The Nuclear Submarine: Riding High', *United States Naval Institute Proceedings*, vol. 101, 2, no. 864, February (1975), pp. 55-62.

8. *Aviation Week and Space Technology*, vol. 117, no. 19, 6 November 1982, p. 13.

9. *Aviation Week and Space Technology*, vol. 117, no. 12, 20 September 1982, p. 13.

10. 'The AEW Helicopter', *International Defense Review*, vol. 16, no. 2 (1983), pp. 188-90.

11. Bill Sweetman and Bill Gunston, *Soviet Air Power* (Salamander Books, London, 1978), p. 89.

12. *International Defense Review*, vol. 16, no. 10, 1983, p. 1482.

13. N. W. Emmott, 'Airborne Torpedoes', *United States Naval Institute Proceedings*, vol. 103, 8, no. 894, August (1977), pp. 47-54.

14. K. R. Singh, *Quarterly Journal of the Institute for Defence Studies and Analyses*, vol. IX, no. 1, July-September (1976), p. 5.

15. Jeffrey Ethell and Alfred Price, *Air War South Atlantic* (Sidgwick and Jackson, London, 1983), p. 219 and Appendix 10.

16. Neville Brown, *Strategic Mobility* (Chatto and Windus for IISS, London, 1963), pp. 117-18.

17. Ezio Bonsignore, 'Vertical Standards on the Way', *Si Vis Pacem Para Bellum*, Year 2, no. 2, February (1978), pp. 45-55.

18. Arthur Hezlet, *Aircraft and Sea Power* (Peter Davies, London, 1970), plate 12.

11 AERIAL MASS DESTRUCTION

The Demise of Strategic Air Deterrence

So at the level of local war, the development of electronics and aeronautics strongly favours the land extending its influence over the sea. Where strategic war is concerned, however, you could fairly say the reverse applies. In 1965, aircraft carriers were withdrawn from US strategic war plans. But by that time the USN was deploying Fleet Ballistic Missile (FBM) submarines equipped with Polaris. They were essentially intended for retaliation against Soviet cities. They complemented land-based programmes already completed or in train.

Over this last quarter of a century, in fact, both Superpowers have maintained a 'strategic triad'. By this is meant an ultimate nuclear deterrent comprised of manned bombers, land-based missiles and sea-based ones. At present France, too, maintains this configuration; and China aspires to do so. In the meantime, however, Britain has come to rely solely on submarines.

Nor is it only in London that analysts can be found who will aver that a flotilla of FBM boats offers a blend of invulnerability and lethality such as to make it alone an entirely adequate instrument of assured retaliation. None the less, in both Moscow and Washington, there are insistent opinions, by no means all of which are blatantly special pleading, in favour of the indefinite continuation of the manned strategic bomber. The debate thus sustained is one of importance in its own right for the future of air power. But it may also offer some indicators as to the outlook for tactical aviation.

Naturally, the mix of arguments has varied. Still, the following is a fair summation of those adduced, in Washington at any rate, these last few years:

> Strategic bombers, unlike Intercontinental Ballistic Missiles (ICBMs) or Submarine-Launched Ballistic Missiles (SLBMs) can be launched, then set in a holding pattern and hopefully then recalled — under firm political control throughout. Therefore, national command authorities do not face as starkly as is the case with ballistic rockets a decision whether or not to go. Besides

which, manned bombers are inherently flexible. They can strike at targets of opportunity, carry out reconnaissance missions, and deliver terminal blows against missiles trapped inside hardened emplacements by the deformations and debris resulting from previous strikes with ICBMs and maybe SLBMs.

Added to which, when strategic bombers are retained by a country with a deep continental interior (i.e., the USA, the USSR or China), it is virtually impossible for an adversary to eliminate them all in a surprise attack. Consider, for example, a Soviet first-strike against bomber bases in the continental United States. Any intercontinental rockets used for this purpose would take some 30 minutes to traverse the north polar region, and would be in line of sight to American surveillance for at least half this interval. Therefore, every bomber being held at ground alert ought to have enough time to take off and enter a holding pattern as envisaged above.[1] Nor could an SLBM attack entirely rule this out.

Lastly, a strategic bomber can have an auxiliary role. It can be used for the swift projection of power into local conflicts in perhaps very distant places. Additionally or alternatively, its basic design can be modified to produce a variant that has this as a prime purpose. Thus one suggestion has been that, taking its airframe from the B-1 design, it would be possible to build a 'Long-Range Combat Aircraft' able to:

> deploy up to 24 Short-Range Attack Missiles (SRAMs) in three weapons bays with eight in each rotary launcher, or 30 cruise missiles, with 16 carried internally and 14 externally in wing pylons, when executing strategic nuclear operations. In a tactical role, it could deploy a maximum number of 142 Mk48 bombs or 84 Mk82 bombs or 15 Captor mines carried internally for use against maritime targets.[2]

Meanwhile, one claim for the B-1B already under development is that it can carry a warload of 75,000 lb maximum or else can carry one of 25,000 lb across 6,000 nautical miles.[3] What the former figure could connote is 150 Mark 82s, the free-falling bombs most favoured of late. Otherwise, it might bear stand-off 'smart' weapons to offset its own lack of tactical agility.

The concept is not without precedents. As Britain's small fleet of Polaris submarines became fully operational around 1972, her squadrons of V-bombers were withdrawn from the strategic nuclear role they had primarily assumed before. Instead, they were reserved for the European and, in due course, maritime environments.

Even before then, however, these planes (which count as medium-range in strategic terms) had recurrently been deployed apropos local situations. Some played a big part in the destruction on the ground of Egypt's air force during the first four days of the 'Suez' campaign of 1956. Some were deployed to Malta in 1961, this in oblique support of an amphibious landing to underwrite newly independent Kuwait against an Iraqi threat. Early in the *konfrontasi* campaign waged by Indonesia — between 1963 and 1966 — against the creation of Malaysia, Britain dispatched a squadron of V-bombers to exercise out of Singapore. This move bespoke a preparedness to strike deep into Indonesia if circumstances so warranted.

Similar commitments have been undertaken by the Strategic Air Command (SAC) of the USAF. Barely three days after the outbreak of the Korean war in 1950, its B-29 Superfortresses were attacking roads leading south into Seoul; and soon these machines and their developmental extension, the B-50s, were to be directed against a diversity of objectives. As a rule, the threat from anti-aircraft guns precluded aircraft so large and unwieldy descending below 20,000 ft over well-defended areas. Nevertheless, the Circular Error Probability (CEP) thus achieved was said to be 150 m with free-fall or gravity bombs and no more than half that with radio-controlled ones. Suffice to note that, throughout this limited war, all the B-29s and B-50s remained integral to SAC's global war plans though pride of place in that regard was gradually assumed by the newer B-36s.

Likewise, a US Senate committee reporting in 1960 on the B-70 Valkyrie advanced strategic bomber then being proposed was to stress its value, too, in theatre war, at least against fixed targets. The previous year, General Thomas S. Power — the then Commander of SAC — had acknowledged before a House committee, albeit with reluctance, that his planes might contribute to a non-nuclear engagement, given several hours' reorientation and technical conversion.[4]

In June 1965 (three months into the full-scale build-up of US ground forces in South Vietnam), B-52 intercontinental bombers flying out of Guam made their first sorties against the Viet Cong.

Within two years, they had dropped nearly 200,000 tons of bombs against targets of various kinds in various parts of Indo-China. Still, their great virtue, as the Pentagon saw it, was an ability to drop exceptionally heavy individual bombs against command posts, hospitals, supply dumps and the like deeply dug within broad stretches of primary jungle. Nevertheless, General Curtis Le May — who had been SAC's first commander — was soon to urge they be held back until 'the time comes . . . to bomb out the port of Haiphong'.[5] He sought thus to resolve any ambiguity about bomber roles by what can be read as an extravagant anticipation of the Linebacker offensives of 1972 (see Chapter 1).

In the aftermath of the Vietnam defeat, the focus shifted towards the part the B-52 might play in sea control. As US Secretary of Defense Rumsfield was to put it in 1976:

> We plan to use available B-52 aircraft to assist in mining operations in a major conventional war. We also feel that such aircraft would be useful in providing anti-ship capabilities with stand-off weapons as well as for some reconnaissance-and-surveillance functions.

Just beforehand, he had stressed that B-52s have 'a capability for quick reaction . . . in the many areas of the world where US naval forces do not normally operate'.[6] Duly, work began on co-ordination between SAC and the US Navy. The fruits of this departure were eventually to be seen in plans to have the B-52s support, in South-West Asia or wherever, the Rapid Deployment Force.

Evidently, too, the Soviet strategic bomber force has long had theatre war as an ancillary role. Not a few of its 600 planes carry air-to-surface missiles suitable for use against ships. The 375 heavy bombers operated by the Soviet navy supplement this capability.

Needless to say, all these elements can also be directed against coastal objectives. For instance, six Tupolev-22Ms were used in direct support of an amphibious exercise against the East German island of Rügen in September 1980.[7] Furthermore, it has been surmised that the gatling cannon each Tupolev-22 and -22M fires to rearward by remote control would be of little value in self-defence but could be employed for the overlap interception of transport or patrol planes.

What has aroused most comment, however, is the placement of racks in two parallel rows beneath the main air ducts of Tupolev-

22s. Some believe these fittings could bear air-to-air ordnance.[8] But others say their configuration is no less suitable for ECM decoys.[9] Meanwhile, there is no evidence of this airframe being adapted to home air defence, say, in replacement of the obsolescent Yak-28P heavy fighter. Conversely, the USAF has considered the possibility of an interceptor version of the B-1, armed with up to 38 Phoenix long-range missiles.[10]

Be all that as it may, however, the main case for the continued use of manned bombers in strategic nuclear deterrence lies open to objection on literally every count. Talk about searching out 'targets of opportunity' looks otiose in relation to the boundless savagery of an all-out nuclear exchange. Then again, a bomber geared to the release of long-range cruise missiles will within hours present the 'launch or cancel' agony of choice almost as acutely as an ICBM site or FBM submarine does. Granted, a bomber relying instead on free-falling bombs can delay the crunch longer, provided its radio links are reliably sustained. However, it is doubtful whether that mode could be viable, in this decade or beyond, in initial operations against either the Soviet or the American homeland.

Nor is doubt on this score diminished by the fact that, long before becoming disabused of the medium-altitude approach to target in the tactical air environment, the USAF was rejecting it in the strategic. Experiments under way as early as 1959 bespoke an awareness that soon the only serious hope of its B-52s drawing close to objectives set in the Soviet heartland would rest on their keeping close to the surface more or less throughout. Likewise in 1963, Britain's Defence White Paper foreshadowed structural modifications — with this end in view — to the Victors and Vulcans of its V-bomber force.

Ambiguity on this score may bedevil the Advanced Technology Bomber (ATB) under development for the USAF as a provisional successor to the B-1. A 'flying-wing' configuration is seen as the most fundamental of its Stealth attributes.[11] Yet a flying wing cannot project a small image to ground-based radar unless it is at a low altitude throughout. Conversely, an ATB needs to be at a good height to conduct the sort of flexible targeting made possible by its exotic sensors.

Furthermore, communication lapses may wreck the command and control of nuclear release. An ominous, if somewhat obscure, precursor is the way half the Luftwaffe bombers converging on Rotterdam on 14 May 1940 failed to receive instructions to turn back issued in the light of a Dutch intimation of surrender. Agreed, the techniques of crisis communication have advanced a lot since then. But so, too, have jamming and deception. Nor dare one ignore the peculiar effects that nuclear bursts have on radios and radars. These are mentioned below in connection with theatre war.

This apart, being able to dispatch bombers into a 'holding pattern' is not so special an option. There are sundry other ways, short of lethal encounter, whereby a nuclear power may express resolve: the evacuation of civilians from key target areas; other emergency measures on the domestic front; the dispatch from harbour of Fleet Ballistic Missile (FBM) submarines; the deployment out of barracks of local war forces; the announcement of economic sanctions; the intensification of strategic surveillance; and, of course, various of these stratagems in various sequences.

In other respects as well, it is only too easy to query or even parody the singular kind of nuclear flexibility the strategic bomber is supposed to proffer. By most of the testable criteria, it is demonstrably less flexible than the ICBM or SLBM. Thus the prospects for the penetration, by machines of this sort, of a well-defended enemy heartland may depend on avenues of advance being opened up for them by extensive bombardment with strategic missiles. They may rest no less heavily on a considerable number of aircraft being committed to each raid. A demonstrative or warning strike by, say, a single strategic missile is far more likely to be feasible than the use thus of a single strategic bomber. Arguably, too, this comparison becomes more pertinent with the prospective emergence of non-nuclear intercontinental bombardment.[12]

What some war games have suggested, however, is that the manned bomber might come into its own towards the further end of the spectrum of all-out nuclear conflict. Once the air defence of the adversary had been thoroughly shattered or dislocated, the vulnerability of the bombers in his skies would be much less. By much the same token, the ability of any such machine to combine strike with reconnaissance would then count for more. So might the facility a heavy warload accords it for attacking either a multiplicity of general targets or else a few unusually tough ones.

Still, to carry the argument this far is to endorse the notion of a global war being fought, with due deliberation, until one side or the other is utterly eliminated. Yet such a notion may be not merely ghoulish but quite unrealistic. After all, when the deterrence debate first got under way, 25 to 30 years ago, there was widespread acceptance that peace through Mutual Assured Destruction (MAD) depended on each Superpower remaining able to deliver about a couple of hundred thermonuclear warheads against the cities of its chief adversary, regardless of the operational circumstances. What this interpretation derived from was the brutal reality that an onslaught of that magnitude could obliterate every urban area of consequence within either territory. Yet were the urban heart of its civilisation to be torn out in this fashion, neither polity could conceivably survive in a recognisable form or, indeed, a coherent form of any kind. So does not this assessment still look valid, never mind that both the USSR and the USA would have thousands of nuclear warheads at the start of a war? In other words, is not enough still sufficient? If so, the penultimate role for the strategic bomber as envisioned above is nothing but an exercise in supererogation.

Next, we ought to examine further the point about bombers compounding the problem that would be faced by an adversary intending to launch a pre-emptive strike. Undeniably, rival strategic panoplies may stand in a relationship one to another such that this consideration could be important. Today, indeed, China could gain a crucial extra measure of immunity from her long-range bomber force, were this not so small and obsolete. Otherwise, her deployed deterrent is still almost entirely in the form of static land-based strategic rockets propelled by 'unstorable' liquid fuel and liquid oxygen. They would take an hour or two to fuel, immediately prior to their actual launch. To which one must add that, because of the elaborated structuring of the tanks and related paraphernalia, no missile site of this kind can be hardened satisfactorily. The implication is that those Chinese strategic rockets could be eliminated wholesale in a sudden attack. Therefore bombers able to 'scramble' from runway alert to a safe distance within several minutes could, in principle, contribute decisively to the stability of Peking's deterrent.

What still has to be spurned, however, is any disposition to see this as an argument for the manned bomber, regardless of the specifics of a situation. One manifestation of this inclination surely was the reaction of SAC to the Cuban missile crisis of 1962. By then, the USAF already had emplaced no fewer than 250 ICBMs. Moreover, most of these were of the new Minuteman solid-fuel type, positioned in well hardened emplacements and eligible for firing at less than a minute's notice. Also, the first of the Polaris boats had been commissioned by the US Navy. The corresponding Soviet panoply, on the other hand, included but 75 ICBMs. Nor did it yet include any SLBMs with ranges above 400 miles.[13] Nor were any strategic missiles on either side then fitted with multiple warheads.

This being so, Strategic Air Command stood in no danger of losing its ICBM echelon in a Soviet pre-emptive strike. Yet its bomber squadrons deployed as though this threat overhung all else. Even before the crisis broke, nearly half the bomber strength of SAC was routinely on a ground alert of 15 minutes, a small airborne alert being maintained as well. Then for the duration of the confrontation, the former was almost doubled while even the latter was brought up to five per cent. The Soviet bomber arm would have needed then to respond thus but not the American.

As things are today, of course, a failure to authorise a bomber alert would leave neither Superpower exposed to being crippled at a stroke. For one thing, each now has the option of firing many of its ICBMs within a minute or two of the alarm being sounded. True, that would be a 'hair-trigger' response, with no room for any second thoughts. All the same, the possibility is one no would-be aggressor could ignore.

However, the case against the USA and the USSR still retaining aircraft as instruments of strategic nuclear power is clinched by the virtual certainty that each will always possess enough long-range missiles of the kinds that are too mobile to eliminate wholesale. Over 20 years ago, the USAF was weighing the desirability of installing Minuteman ICBMs in railroad cars. Hardly surprisingly then, overland mobility can readily be incorporated in contemporary specifications for ICBMs. Witness the American MX and Midgetman programmes. Witness, too, the Soviet SS-20: a system that may fly a good 4,000 miles with a light warhead.[14]

Above all, however, the installation of many hundreds of SLBMs today ensures that either Superpower could always respond massively to any fell swoop against its home territory. In other words, underwater mobility affords a deeply credible guarantee of second strike, regardless of strategic bombers.

Nor is this state of affairs at all likely ever to be overturned by technological change. As often as not, discussion of whether such an upset could take place revolves, by implication at least, around a comparison with the status of Anti-Submarine Warfare (ASW) towards the end of the Battle of the Atlantic (1939–45). However, the setting for a strategic nuclear exchange would assist the submarine more. An FBM is not concerned, unlike the U-Boats of World War II, with closing to action with adversary craft. Instead, it has a duty to avoid them. In this aim it is assisted by the attributes conferred by nuclear-propulsion and the novel hull-designs associated with this.

Besides, even in May 1943 — a month of decisive victory for the Allied escort forces on North Atlantic stations — the average attrition of the U-Boat fleet at sea was barely one per cent per day. Yet a pre-emptive first strike against an FBM force, extremely well-dispersed across the high seas and under polar ice, would have to achieve close to 100 per cent success in well under half an hour. Even in terms of that simple arithmetic, the required improvement in ASW effectiveness is ten thousand fold![15]

But the question that remains is whether the deadlock here described might not eventually be broken by advances in Ballistic Missile Defence (BMD). By this is now meant, first and foremost, BMD in the form of lethal beams (laser or maybe subatomic particles) directed from battle stations orbiting in near space. So what are the prospects for the accelerated research and development in this sphere announced by President Reagan in his 'Star Wars' speech of March 1983? To take the alternative tack, could advances therein by the USSR oblige her adversaries to continue with strategic weapons (bombers or cruise-missiles) that never ascend above the protective blanket of the earth's atmosphere?

Any assessment must hinge on certain aspects of Quantum Theory, a corpus of thought fundamental to modern physics. Among them is the tenet that the wave forms assumed by projected beams (whether composed of energy or particles) always diverge and hence decrease in intensity as distance increases. The longer the wavelength, the greater the divergence. The narrower the aperture of the beam emitter, the greater the divergence. Accordingly, the most serious obstacles to the successful development of the proposed battle stations arise from the constraints on aperture width when an emitter is placed in orbit and from the difficulties in developing shorter wavelength lasers. Also, similar particles within a beam sometimes repel each other markedly because they bear electrical charges of the same sign. Inevitably, too, there are severe limits on the establishment of power sources in orbit; and major problems in training a beam steadily on an ascending missile.[16]

All of which points towards the conclusion that no network of

BMD battle-stations can be established in near space, at anything like a tolerable cost, for a good 20 years. For this seems the time required for X-ray lasering to advance enough to be applied to this end. What is more, advantages that might accrue from projecting beams at such short wavelengths could only be realised if the number of stations in orbit were kept correspondingly low. Yet the fewer there were, the more susceptible the network as a whole would be to pre-emptive interception by anti-satellite weapons, systems the USSR has been evolving since 1968 and which the USA now urgently works on.

Besides, apprehension that the MAD stalemate might be overturned by space-based BMD would not ipso facto justify perseverance with the manned strategic bomber. By such a time, the year 2010 or thereabouts, ultra long-range cruise missiles launched from the surface are likely to be available. Reportedly, intensive work is under way along these lines in the USSR as well as the USA.[17] None the less, a new Soviet heavy bomber codenamed Blackjack by NATO) was first flight-tested in 1982. Even in the interval before the advent of ultra long-range 'cruise', however, it is hard to see that such machines have a contribution to make to strategic nuclear war.

Theatre Confrontation

But the next question must be whether the above objections to the manned aircraft apply with theatre nuclear deterrence. Air staffs have tended to say they do not. The most recent instance is the French decision to keep 18 Mirage IVs in service until 1996 but arm them with tactical nuclear stand-off weapons.[18] So how crucial is the kind of flexibility the warplane proffers to the graduation of a theatre nuclear response? And how far will its merits in that regard be overshadowed by its extreme exposure to nuclear obliteration on base?

By the late 1950s, agonised endeavours were being made, especially in the USA, to work out guidelines for the graduated escalation of nuclear conflict in Europe. In various studies conducted then, military airfields were deemed more appropriate as nuclear targets than other enemy assets at comparable distances behind the front. Within a few years, however, the notion that escalation could be controlled by means of pre-defined 'firebreaks' was to be all but abandoned.

Then during the Kennedy administration, the stockpile of theatre nuclear warheads in NATO Europe was doubled to the level of 6,000 it was to stay at until just recently. Moreover, it was apparent that many hundreds of these devices were intended for delivery by warplanes. What is more, their great diversity included a high proportion of the higher yield warheads: that is to say, a high proportion of those in the megaton range — of the order of a million tons of high explosive. The aggregate explosive yield of the 6,000 was put at 400 million tons of TNT equivalent.

Likewise, several hundred nuclear warheads are integral to the US garrison in the Republic of Korea (ROK). Most are held by the US Army. Presumably, however, their number has lately decreased in association with the cutback in US ground troops announced in 1977. The remainder are assigned to the F-4s of the USAF. These are normally based at two airfields, Kunsom and Osan.[19] More speculative though by no means implausible was the judgement made in 1981 that Israel had thus far fabricated 30 nuclear warheads, some of which were earmarked for aerial delivery.[20]

Among the options available is that of using aircraft to make the first selective application of nuclear firepower. One aim would be to gain some military advantage, the destruction of a key bridge or whatever. Another would be to make a psychological impact, in order to secure a 'pause' in dynamic warfare. This could afford all concerned a chance to pull back from the brink. None the less, aircraft were mainly seen within NATO as presenting the threat of overkill: as the means of inhibiting massive escalation by the enemy within the European theatre through their own ability to retaliate massively and in depth. In short, they were the putative guarantors of MAD, at the European level and likewise, it seems, the Korean. By the same token, they coupled theatre nuclear deterrence to global.

Yet herein lay a dangerous paradox. What were supposed to be a prime deterrent against nuclear or chemical escalation could just as well prove a positive invitation to it. For the planes in question were themselves located around a limited number of well-known runways. What then of the bold claim of British officialdom that in the United Kingdom, at any rate, ground-based surveillance radar would provide four minutes' warning of any missile attack from Warsaw Treaty territory; and that V-Bombers held at Quick Reaction Alert (QRA) at the ends of runways could 'scramble' to a safe distance within that time?

One difficulty about such fine reckoning was the estimate of four minutes' grace. It derived from the assumption that rockets travelling the thousand miles or so from Eastern Europe would average three miles a second during the long exo-atmospheric phase between burn-out and descent to target. But that expectation was geared to the further one that the fuel consumption in each rocket had to be kept to a minimum in pursuance of the best combination of warload and range. Suppose warhead size and rocket performance were such that fuel conservation was not an overriding imperative. Then the planned speed on burn-out might be raised to four or five miles per second, and the flight trajectory flattened. Together, these changes would much foreshorten the warning time afforded the RAF. So might radar-jamming. So would the use of SLBMs against the bomber stations.

Then again, every key airfield could be bracketed by the near-to-simultaneous impact of several missiles. After all, no fewer than 600 Medium- and Intermediate-Range Ballistic Missiles were deployed within the western borders of the USSR in just the two years between 1961 and 1963. At all events, there was never a prospect of anything like all the V-Bomber force surviving a comprehensive attack against every field it may have dispersed to, in flights of four planes apiece. Nor might any bombers that did survive a missile strike then be able to form up in the sort of co-ordination required for penetration of Soviet territory.

Nor does continuous airborne alert ever look like an adequate remedy. Nobody believes it could ever be kept at more than a tiny fraction of total strength, year in and year out. Nor might any very adequate level be sustainable for more than the first several days of a warlike confrontation.

In the light of which, it is truly remarkable that, before the said weapon was cancelled by the Kennedy administration in 1962, the British government hoped to keep its deterrent up-to-date by equipping the V-Bombers with Skybolt, a longish-range USAF missile. Almost everything that has just been said about exposure to pre-emptive destruction would have applied irrespective of whether these planes carried Skybolt. The one exception is perhaps that planes loaded with free-falling bombs have to penetrate Soviet airspace more deeply and so have a more imperative need to form the co-ordinative patterns alluded to above. Otherwise we see here yet another sobering example of an unhelpful disposition to play down the problem of dependence on runways.

Inevitably, the dilemmas thus left unresolved were and are even more acute apropos air bases nearer the Soviet bloc. In fact, this is why the nuclear doctrine of the NATO air arms became, in the early 1960s, 'trip-wire' with a vengeance. Certain planes and crews stood at the ends of runways on Quick Reaction Alert (QRA), ready to launch full-scale nuclear retaliation just as soon as a threat became active.

Again one must question whether such a solution could ever have worked without pre-emption. In any case, it was utterly at variance with notions about graduated recourse to nuclear firepower. Duly, the air arms had to be brought into line when, upon French withdrawal from the military side of the alliance in 1967, NATO Council became free formally to adopt a strategy of 'flexible response'.

Yet even then, no attempt was made to address more fundamentally the contradictions inherent in reliance on aircraft for this nuclear role. Nor have they been properly addressed since, despite being all too evident to certain of NATO's most senior commanders. Yet they were rendered more acute than ever in January 1984 by the USSR's introduction into Eastern Germany of SS-22 missiles. From there, these weapons could be directed against fixed targets within a broad arc from the Moray Firth to the Pyrenees. Despite much political commentary in a contrary sense, one is bound to say that road-mobile GLCMs and Pershing 2s are far less threatened by these weapons than are machines that roar down runways.

In the above discussion of strategic bombers, there was no need to dwell on the more esoteric problems presented by the nuclear environment. Indeed, to have done so would have been otiose because (a) the outcome of total war would hardly rest on nice calculation; and (b) the drawbacks of the manned bomber in such a context go deeper than the specifics of performance. Much the same might be said, of course, in relation to theatre engagements in Europe or North-East Asia. All the same, theatre nuclear release is always a rather special case. After all, it might have to be embarked on very urgently and yet remain tightly delimited throughout. So it may be pertinent here to look at the operational aspects.

Not that doing so will much assuage doubts about the part manned aircraft can play. One type of warhead under consideration for the Pershing 2 is a 4,000 lb Earth Penetrator able to produce a sub-kiloton nuclear detonation having driven itself, by means of a rocket charge, 50 or 100 feet into the ground. The estimate is that 500 men would have to work for three days to repair at all adequately the heave-and-shatter damage this could cause, say, to a runway.

A further complication apropos the defence of airfields or other rearward targets is that the interception of nuclear-armed aircraft or tactical missiles can itself cause extensive collateral damage. Among the hazards is that of inducing the detonation (maybe by 'dead man's fuses') of the high-yield warheads an intruder may contain. That could easily cause spontaneous ignitions of combustible materials across slant ranges of tens of kilometres.

Besides which, one side or the other might feel tempted or obliged to arm its interceptor missiles with nuclear warheads. The low-kiloton explosions these would produce would be unlikely to cause general combustion across more than a few thousand metres. Given a clear atmosphere, on the other hand, they could probably induce retinal burns more or less to the horizon. To date, however, there has been no great inclination to introduce nuclear warheads into SAM sites. So far as the West is concerned, in fact, the only land-based SAM systems that have been accorded this nuclear alternative are two US ones (the Nike Hercules introduced in 1958; and perhaps its chief replacement, the Patriot) plus a British, the Bloodhound 2, which came into service in 1962.

Meanwhile, the use of nuclear warheads in air-to-air missiles has been further inhibited by the difficulty that miniaturisation presents. Therefore the only such weapons the West has ever fitted thus are those allocated to NORAD's now obsolescent F-106 force, the Genie and the AIM-26 Falcon. They weigh 840 lb and 200 lb respectively; and their nuclear charges have a yield of between 0.5 and 1.5 kilotons. Of the eight or nine varieties of air-to-air missile the USSR has thus far brought to operational service, only one — the AA-6 used on the MiG-25 — is thought able to pack a nuclear charge. At an estimated 700 lb, the AA-6 weighs enough. Still more to the point, its warhead capsule closely resembles the shell fired by the now obsolete S-23, a 180 mm gun/howitzer. Western analysts (who used to identify this ordnance as the 203 mm M-1955) regularly assumed that it could fire a kiloton charge. Then again, suspect though the Khrushchev memoirs are on account of KGB involvement, it is of interest that, in the course of an otherwise dismissive reference, they, too, credited the S-23 with a nuclear capability.[21]

At 45,000 ft, say, the atmosphere is only about one twelfth as dense as at sea level. Blast waves, nuclear or otherwise, are weakened correspondingly. Moreover, the absence of reflection means that an airframe is always better able to ride out a blast in flight than if on the ground. Nevertheless, even at high altitude a kiloton burst would be liable to wreck a heavy bomber, say, 500 ft away; and lower down, it would across two or three times that distance. In addition, a warplane in flight is peculiarly susceptible to heat and to Electro-Magnetic Pulses (EMP).[22] The latter are the emissions (electromagnetic and particularate) distinctively caused by nuclear bursts. They may severely compromise semi-conductors, circuitry, actuation devices and so on across scores or even hundreds of miles. Nor should one forget that aeroplanes aloft are often in a line-of-sight relationship to explosions on the surface as well as to those in the sky. Granted, their resistance to radiative nuclear effects can be built up, as witness work already done to this end on the B-1. But the problem is manifold.

As a rule, intruding aircraft would stand to benefit tactically from the extensive and rather persistent black-outs that may be caused, over a wide range of radio and radar frequencies, by nuclear bursts in the ionosphere: the electrically active zone of the high atmosphere. However, the improved prospects for sortie survival would be overshadowed by the likely interference with wireless links. That could compromise drastically the control by higher authority of nuclear release. Yet for such control to be effective would be even more imperative than with, say, nuclear-armed artillery guns. One reason for saying this is that aircraft travel so far so fast. Another surely is that a cockpit is hardly an ideal environment for nuclear deliberation without external guidance.

Even so, one cannot escape the simple fact that to use a warplane for nuclear delivery is to accept that, at least in the final stage of the approach to target, executive control will be transferred to two crew members or even just the one. Granted, we are told that under one relatively refined arrangement — that for the Mirage IV — nuclear arming by the authorities:

> takes place in flight. In order to drop the bomb from the aircraft, one needs a 'key' which is given only during the mission, and the code for which the pilot does not know. The code is only given during the flight as well. In these planes, the firing circuits also

involve cooperation between two individuals, the pilot and the navigator.[23]

All the same, a majority of those types of tactical aircraft rated nuclear-capable are only single-seat. Among them are the other three French: the Mirage IIIE, Jaguar and Etendard. They also include the several Soviet ones (the MiG-21, -23 and -27; and the Sukhoi-7, -17 and -24) as well as such American or British models as the F-16 or the AV-8B Harrier.

Indeed, arming may often take place on the ground. There it may involve the precise designation of explosive efficiency and hence yield. Also the routine may include the unlocking by higher authority, through remote control, of an electronic device called in American parlance a Permissive Action Link (PAL). Alternatively, the PAL may be activated in flight, much as with the Mirage IV. However, no arrangement of this kind can offer anything approaching an absolute guarantee that an erring or aberrant pilot will not direct a nuclear warhead against the wrong target or, of course, no target at all. Nor does the provision of a second crew member entirely exclude such possibilities. Nor ought one to forget that a plane is much more likely to be shot down short of its target area than is, say, a Pershing 2. Nor that we know little or nothing of Soviet arrangements, relevant though these could be to our own tactical philosophy and arms control policy.

Prior to actual nuclear arming, the tactical commander would probably have to switch tactical strike aircraft from the non-nuclear roles they would otherwise perform. At one time, reloading and other technical adaptations (the switching of firing circuits etc.) could take twenty-four hours or more. Nowadays, it is more a matter of several hours, at least within NATO. But even that much delay would be bound to accentuate a conflict of priorities that would be serious in any case.

A further snag is that it is not feasible to have an aircraft (be it monoplane or rotorcraft) sealed against radioactive fallout the way a battle tank can be. By the same token, the former cannot be properly insulated against lethal chemicals. What can be said is that aircraft are able to pass well above contaminated ground or quickly through clouds of contamination.

What cannot be gainsaid, however, is that the open plan and static character of airfields exposes them to toxicity no less than to explosive force. In 1969, this author was advised of the extreme concern of the British air staff over the vulnerability to chemical attack of its airfields in Germany.[24] Just ten years later, and very much in line with the British tradition of airfield defence, the RAF launched a big new programme for the passive protection of these stations against chemical attack. Among its key features have been making revetments airtight and the provision of special clothing for ground staff: measures which, needless to say, afford some protection against fallout as well.

Unfortunately, however, such precautions are prone to cause a sharp drop in working efficiency. Moreover, they could never ensure that aircrews and ground staff will be shielded from toxicity completely. To make a revetment airtight may be literally worse than useless if doors are open just as it is enveloped in wisps of contaminated air; and, of course, the same applies to a cockpit. Another particular hazard is absorption of contaminants by the coating of special paint usually integral to Stealth technology.[25]

Warplanes are also earmarked to play a prominent part in any recourse by NATO to chemical weapons. Here again, however, one has to question their suitability compared with such alternatives as artillery guns, the Pershing 2 or the SS-20.

What might perhaps be said is that aircraft are the most suitable, or maybe least unsuitable, for initial nuclear release at sea. But this would be hard to argue definitively for the simple reason that the whole debate about nuclear recourse at sea has remained so vacuous.[26] Contingencies are peculiarly unpredictable. The environment is so open-ended. Naval staffs have been loth to indicate operational priorities. There has been too little discussion in the public domain.

References

1. Colin S. Gray, Adelphi Paper 140, *The Future of Land-Based Missile Forces* (The International Institute for Strategic Studies, London, 1977), p. 5.
2. Jacquelyn K. Davis and Robert L. Pfaltzgraff, *Power Projection and the Long-Range Combat Aircraft* (The Institute for Foreign Policy Analysis, Cambridge in Massachusetts, 1981), p. 30.
3. *Aviation Week and Space Technology*, vol. 120, no. 9, 27 February 1984, p. 17.
4. Neville Brown, *Strategic Mobility* (Chatto and Windus, London, 1963), pp. 151–2.
5. Curtis E. Le May, *America is in Danger* (Funk and Wagnalls, New York, 1968), p. 261.

6. *Testimony on Department of Defense Appropriations 1977, Hearings before a subcommittee of the House Committee on Appropriations* (US Government Printing Office, Washington DC, 1976), pp. 388 and 797.
7. *Defense and Foreign Affairs Daily*, vol. 10, no. 104, 1 June 1981, p. 1.
8. *Aviation and Marine International*, 79, December (1980), p. 26.
9. Bill Sweetman and Bill Gunston, *Soviet Air Power* (Salamander, London, 1978), p. 170.
10. *Aviation Week and Space Technology*, vol. 117, no. 9, 30 August 1982, p. 11.
11. 'Decision on B-1', *Aviation Week and Space Technology*, vol. 122, no. 12, 25 March 1985, p. 13.
12. Carl H. Builder, Adelphi Paper 200, *The Prospects and Implications of Non-Nuclear Means for Strategic Conflict* (The International Institute for Strategic Studies, London, 1985).
13. *Military Balance, 1962/3* (The International Institute for Strategic Studies, London, 1962), pp. 3, 9 and 28.
14. *Christian Science Monitor*, 5 September 1984.
15. See the author's *The Unbreakable Nuclear Stalemate* (The Council for Arms Control, London and Windsor, 1982), p. 5.
16. See the paper presented by the author during the 6th National Conference on Quantum Electronics at the University of Sussex in September 1983. It was reproduced in *Navy International*, vol. 88, no. 11, November (1983), pp. 676–81.
17. 'Soviets Test New Cruise Missiles', *Aviation Week and Space Technology*, vol. 120, no. 1, 2 January 1984, pp. 14–16.
18. Paul Jackson's 'Deterrent At The Crossroads', *Armed Forces*, vol. 4, no. 3, March (1985), pp. 101–2.
19. S. E. Johnson and J. A. Yager, *The Military Equation in Northeast Asia* (The Brookings Institution, Washington DC, 1979), p. 47.
20. Michael Dunn, *Defense and Foreign Affairs Daily*, vol. 10, no. 116, 17 June 1981, p. 1.
21. Nikita Khrushchev, *Khrushchev Remembers: The Last Testament*, ed. and trans. Strobe Talbott (Little, Brown and Co., Boston, 1974), pp. 52–3.
22. Henry A. Kissinger, *Nuclear Weapons and Foreign Policy* (Harper, New York, 1957), ch. 4.
23. *Tactical Nuclear Weapons: European Perspectives* (Taylor and Francis for Stockholm International Peace Research Institute, London, 1978), p. 68.
24. For an allusion to this advice, see Neville Brown, *British Arms and Strategy, 1970–80* (Royal United Services Institute for Defence Studies, London, 1969), p. 30.
25. *Flight International*, vol. 119, no. 3748, 1 March 1981, p. 617.
26. Eric Grove, 'Nuclear Weapons in Surface Navies — More Trouble than They Are Worth', *Defense Analysis*, vol. 1, no. 2 (1985), pp. 135–6.

PART FOUR

A PREDICTIVE OVERVIEW

12 COSTS VERSUS BENEFITS

Investment and Operation

Some sense of the balance between the economic burden imposed by air power and the military benefits or dividends to be derived from its employment will properly play a part in the debate about its future. However, few people would deny these days that the benefits cannot be measured, let alone predicted, with arithmetic exactitude. What ought now to be stressed is that the same applies to costs as well. Take, for example, procurement. The stated terms of sales contracts (domestic or abroad) are often treated as though they were definitive guides. Yet rarely has one been told what pricing policy has been adopted in a given instance: marginal cost; average cost; or whatever. Very generally, too, nothing is said about the inclusion of spare parts.

The yardstick analysts usually favour is an aircraft's 'flyaway cost'. Yet this excludes a due fraction of the expenditure incurred on research and development. Nor does it allow for spare parts or specific support equipment. Nor may it always include the warload. To which must be added the basic objection that a single such statistic ignores the way the incremental outlay per aircraft (i.e. the marginal cost) will tend to fall as the total coming off the assembly line rises. Take, for instance, the US production, during World War II, of almost 16,000 P-47 Thunderbolt fighters. Whereas the first 733 had an average flyaway cost of $113,250 at 1944 prices, the remainder had $83,000. Indeed, the cost of a US fighter of that generation was unlikely to level out until the 20,000 mark had been passed.[1] Even today, 3,000–5,000 would be a representative bracket for this levelling for warplanes and also various missiles.

Then there are the difficulties inherent in making transnational comparisons. These are always aggravated by contrasting input patterns, notably *vis-à-vis* the balance between labour and capital. But these days they emanate mainly from fluctuations in the rates of currency exchange, together with the intricate interplay between these and relative internal prices. Likewise, inflation complicates comparisons over time. Between 1945 and 1967, the US consumer price level rose by 82 per cent.[2] Then in eight years it was to nearly double and in eighteen a little more than treble (see Appendix B).

228 *Costs versus Benefits*

The year 1975 has here been taken as the main base year because (a) it must be close to the median date for the data employed; and (b) it avoids the acute instability of the international exchanges in the interim.

Not that monetary trends could mask at all completely those in real procurement costs. A prominent theme in the literature has been how fast the price of warplanes has risen compared with those of land weapons. With tanks, in particular, the divergence is pronounced. In 1943, the cost of a main battle tank (MBT) — a Grant or Sherman, say — was $55,000 or $160,000 at 1975 prices. In 1978, a comparable figure for a modern MBT was $550,000.[3] The comparable figure for a large self-propelled piece of field artillery would rather exceed a million dollars. That for its towed counterpart in World War II might have been an order of magnitude less.

One reason for a more marked escalation in respect of advanced warplanes is that, unlike the roads and tracks used by land vehicles, runways can often be made much larger and stronger. Alternatively, planes may cling to them less. Another reason is, of course, the scope afforded exotic technologies by the aerial ambience. The mean cost of a P-47 would have been roughly $300,000 in 1975. Obversely, the first version of the F-4 Phantom (which entered service in 1961) would then have cost $3,800,000 and a later version up to $5,750,000, this despite the eventual production by McDonnell-Douglas of over 5,000 Phantoms of one kind or another. Moreover, the flyaway cost of an F-14 Tomcat, a fighter in USN service since 1975, has been $17,000,000 or over 50 times the P-47s.[4] Other flyaway costs for representative aircraft were given by the USAF Director of Management Statistics early in 1976, using the money values then prevailing. A production run of 120 Rockwell B-1 bombers would yield an estimated figure per aircraft of $57 million; and one of twice that, $47 million. An omnibus estimate for the E-3A Sentry was also $47 million while for the austere A-10 Thunderbolt 2, it was $3,500,000.[5]

What then of the exclusion from flyaway cost of both support equipment and initial spares? American practice is to include these ancillaries within 'unit procurement costs'. With the A-10, they add a mere $600,000. But with the E-3A (geared, as it is, to elaborate electronics), the extra is $14,000,000. Nor can one ever ignore the development budget. With recent US fighters, this has been around $1,500 million. For the Advanced Technology Bomber it is expected to top $4 billion.[6] Even the A-10, its austerity notwithstanding, incurred $500 million.

But might certain factors together make the capital costs of aircraft much lower in the Soviet case? Among them could be less weight and sophistication. So could a tendency for lower wages to more than offset low labour productivity. A production run of a good 10,000 (all versions included) will also apply in the particular case of the MiG-21.

Even so, one has to discount the more extreme renderings of this particular theme. Among them have been that a MiG-21 may lately have cost only $1,600,000 or a Sukhoi-7 (Su-7) $1,800,000; and even a MiG-23 or Su-20 a mere $2,000,000.[7,8] These inferences have been drawn from information about sales abroad without due regard to the caveats entered above. One specific factor is Moscow's insistence on keeping the rouble undervalued. Also, her politicians may have been disposed to subsidise arms exports heavily. At all events, no credibility attaches to a differential between the MiG-21 and the MiG-23 of but $400,000 or between the Su-7 and the Su-20 of a mere $200,000.

Military airfields also represent heavy investment. Thus the buildings and other fixed assets in the RAF's front-line station at Bruggen in West Germany have been said to top $500 million, as against about $425 million for the aggregate flyaway costs of the 60 Jaguars it used to operate.[9] Much worse, though, is the cost of basing at sea. A Carrier Vessel Attack Nuclear (CVAN) costs three billion dollars to build — a good $30 million for every aircraft borne. To which must be added the capital costs of a high proportion of the many other ships in a CVAN task force. So, too, must some quotient of the infrastructure ashore; jetties, barracks, airfields and so on. Allowance must then be made for ship rotation. In the ultimate, too, a disconcerting proportion of all air power afloat may have to be reserved specifically for the defence of the parent task force. Nor does recourse to mini-carriers ease the financial burden per strike plane. How could it in the light of what has already been said about tonnage and aerial complement?

Ammunition is a category which might be treated by an economic purist as consumption rather than capital. Since little is consumed in peacetime, however, it is best dealt with here. Let us turn first to ordnance employed by aircraft against surface targets. A 1,000 kg unguided and unitary bomb cost $1,000 in 1975 in the USA, as compared with $100 for a 16 kg 105 mm artillery shell. That comparison as such is favourable to the shell, given that the cube-root law applies (see Chapter 4). On the other hand, the cost of a 1,000 kg cluster bomb with 1,000 bomblets inside it might only rise to several thousand dollars. But with that kind of configuration, cube roots do not work against air delivery.

A teleguided Maverick air-to-surface missile may cost $22,000 on a production run of 30,000, while a Harpoon anti-shipping missile has been said to cost $440,000 on a run of but 785.[10] Then when a deflation of 19 per cent is applied to 1978 quoted prices, the 1975 cost of 1,000 rounds of 12.7 mm belted ammunition becomes $1,175; and that of a single round for a 20 mm cannon, $8.2.[11]. At this point, however, one is starting to consider munitions that can also be used against aircraft. So the implication that nearly a couple of million 20 mm cannon shells, say, could be purchased for the flyaway price of a single Tomcat is disconcerting.

Expenditures on individual warplanes look similarly immoderate when set against those of the systems (and especially the lighter guns) that shoot from the surface at them. The 1975 cost of a singly mounted 12.7 mm heavy machine gun would be $4,500. That of an M-167 (the towed version of the Vulcan six-barrel 20 mm anti-aircraft gun) would be $245,000. True, the M-163, a tracked and self-propelled Vulcan, would cost $645,000. Yet even this implies a fiscal equivalence of 26 Vulcan guns to one Tomcat. Finally, the Sergeant York divisional air-defence system (twin 40 mm guns on an M-48 tank chassis) would have cost just over $4,000,000 on a production run of 600. Apparently, this average excludes the several hundred million being spent on development when this system was cancelled (for specific technical reasons) in the summer of 1985. A SAM launcher of comparable role would tend to cost rather less. On the other hand, its guided missiles would be individually expensive.

Still, even the upper end of this spectrum compares favourably, in these naive terms at least, with the outlays on warplanes. As much is even more apparent with those ground vehicles liable to be prime targets for air strikes. A British light tank, the eight tonne Scorpion, sells for a modest $125,000. That fact gains significance from the likelihood that the Scorpion may anticipate the representative tank designs of the late 1990s more accurately than do the 40 to 60 tonne Main Battle Tanks (MBT) of today (see Chapter 5).

Admittedly, tanks and the like will usually present themselves in some number: in convoy or park or open formation. Granted as well, the use of spread weapons (not least the minelets) enables multiple armoured targets to be engaged more readily than before. Then again, aircraft loss rates will vary markedly according to circumstances. Even so, it is fair to observe that something like a Scorpion will rarely be a worthwhile individual target for a far more expensive warplane. Moreover, the disparity in capital commitment is even starker with machines like the M-113: the tracked Armoured Personnel Carrier (APC) the US has produced at $60,000 apiece. On the other hand, there may be as many as eleven infanteers inside a M-113.

Just about the only other sphere of weapons development in which, over three or four decades, the escalation of real costs has more or less matched that of warplanes is that covering such categories of warship as destroyers, frigates and, of course, aircraft carriers. Here as with the planes, big increases in sheer size have taken place. Once again, this is partly because the craft in question are not tightly confined, the way land vehicles are, by such parameters as the width of village streets. With the overall trend at sea, however, have come reformulations of mission concept that have bordered on the chaotic. Accordingly, comparisons of past and present cost cannot be drawn at all exactly. But what is safe to say is that the growth in warship size has now peaked out (see Chapter 10).

Clearly, everything just mentioned is in oblique competition with manned aircraft *vis-à-vis* the delivery of firepower or the conduct of surveillance. But with certain instruments of war, not least of strategic nuclear war, the competition is more direct. Again, one has to say that the choices must never solely hinge on trite economics. In any case, the introduction into strategic missiles of Multiple Independently-targetable Re-entry Vehicles (MIRV) has not given these systems an edge over the manned bomber, on a straight comparison of capital cost, as clear as could have been expected. This is largely because of the effort undertaken these days to guard such systems against pre-emptive destruction, either by mobility or else through site-hardening. Thus the full sliced cost of a Trident is in the $150 to $250 million bracket. Therefore the case in favour of this system, as against the B-1 bomber, has to be clinched by adducing operational (and, in most cases, incommensurable) factors: command and control; penetration prospects; strategic scenarios; the course a nuclear war might take; multiple targeting; flexible targeting; ancillary missions; running costs; and the interaction with adversary policy.

232 Costs versus Benefits

What then of the advent of strategic cruise missiles? Early in 1977, a USAF spokesman testified that the flyaway price of an Air-launched Cruise Missile (ALCM), warhead excluded, would be $900,000 at the money values then obtaining. Meanwhile, naval testimony gave the flyaway cost of a sea-launched ALCM-B Tomahawk as $690,000. Then again, the Pentagon was expecting the development budget for this cruise-missile generation to be a modest $1,050 million, this to be spread across a production run of perhaps 3,500 of the air-launched and sea-launched varieties. So in those elementary terms, at any rate, the cost advantage over deep-penetration manned bombers looks considerable in the context of strategic war. It would become dramatic were one to give credence to another opinion expressed by Richard L. Garwin, the IBM analyst. He said that, on a production run of 100,000, the procurement cost of an ALCM need not exceed $100,000.[12] With respect, however, that suggests economies of scale much greater than manned warplanes have ever yielded.

Besides, one official view has been that to ensure acceptably high rates of penetration by the ALCM-B (through the on-board installation of ECM plus a reactive manoeuvring device) might well be to multiply the flyaway cost as much as six times. Nor can one ignore collateral investment in aerial launch platforms. Adapting a Boeing 747, say, to the operational deployment of 72 cruise missiles could be to treble its unit cost to over $90 million.[13] On the other hand, as and when such devices are ground-launched, they require little in the way of support facilities.

However, another comparison to draw is this. The outlay on a cruise missile tends to compare unfavourably with that on an equivalent ballistic model. This is despite the fact that, as range increases, the latter incurs an ever greater penalty from having to carry its own oxidiser instead of 'breathing' air. The basic cost of a Lance ballistic missile (which travels up to 120 km, surface to surface) was originally put at $115,000; and this figure is officially expected to halve, as follow-on versions are mass-produced.[14] Even a Pershing 2 missile is understood to cost, at 1976 prices and with a two ton non-nuclear warhead, only $1,000,000.[15]

Evidently, the costings ascertainable for cruise missiles can serve as a rough guide (albeit one on the low side) to those for unmanned reconnaissance vehicles of similar sizes and aptitudes. Additionally, a more specific comparison might usefully be made between the Aquila (see Chapter 7) and the RF-4, a version of the

Phantom with special reconnaissance systems installed. One of these is the AAD-5 downward-looking infra-red sensor with film recorder. Altogether, this cost $250,000 in 1977. Another is the ARN-101, a Loran-C system for exact self-location. Then there is a three metre pod called Pave Tack. It contains electro-optical and Forward-Looking Infra-Red (FLIR) sensors, along with a 'laser target-designator'. It cost another $750,000. In addition, a threat detector plus other listening devices are borne; and maybe a Sideways-Looking Airborne Radar (SLAR) as well.

Altogether then, a versatile package costing $1,500,000 is being carried in a plane otherwise costing $4,000,000. In itself, that proportion is not intolerable. But more worrisome is the disparity in weight between surveillance equipment and vehicle. The worry is accentuated if one envisages tactical situations so fraught that the two-man crew can itself contribute little to the reconnaissance task. Thus the payload just referred to weighs barely 1,300 lb which represents but one part in 40 of the loaded weight (or one part in 23 of the unfuelled weight) of an RF-4. At 38 lb, in contrast, the surveillance package in the Aquila weighs nearly a third of what the original version of that machine does. Lately, however, the Aquila's overall weight has been allowed to rise by two-thirds in order to accommodate an anti-jamming device and night-vision equipment. Reportedly, too, the cost of a complete Aquila will therefore inflate eight-fold to $800,000.[16] Still, the 1985 costs of RPVs as a genre mostly lie between $750,000 and a mere $25,000. The figure for a given model depends, in the main, on range, payload, production run, and whether it is meant to be reusable.[17]

Therefore, the divergences between manned and unmanned are already unfavourable to the former with reference to missions that are short-range, in pursuance of aims not too complex, and in an ambience comparatively benign. Manifestly, too, RPVs stand to profit more from the progressive miniaturisation of components, especially electronic ones.

As has just been intimated, however, the balance of capital cost will be more favourable to the manned aircraft where really deep and extensive reconnaissance is called for. So will it when the task in hand is not probing or surveillance but active and diversified Electronic Warfare. Take the EF-111A, the EW variant of the F-111. Its EW payload of 4,000 kg comprises a tenth of that plane's all-up weight; and makes up half the flyaway cost of $20,000,000. Nor should anybody discount the great contribution the dexterity of an on-board human operator can make in electronic combat.

Helicopters, too, appear in remotely piloted forms or as drones. But their more familiar manned variants must also be assessed as alternatives to manned monoplanes. As a rule, they seem a less expensive buy than might have been assumed, given their complexity and the demands imposed by vertical lift and static hover. Yet once again, coherent discussion is bedevilled by the ambiguities that surround export prices and so on. But several million dollars seems a typical flyaway cost for a helicopter of around 10,000 kg all-up weight.

Overall, the procurement cost of manned aircraft (relative to that of antidotes or alternatives) does not yet seem so high as to rule out their contribution ipso facto. Still, a lot may hinge on what the future holds for the procurement cost of the tactical monoplane. The 1974 Brookings study cited earlier opined that, on the trends then current, it could 'become entirely plausible' that 'the replacements for the F-15 and the F-14 that can be anticipated *in the mid-1980s* . . . will cost $100 million or more apiece (plus the cost of research and development and accessories, and an inflationary mark-up)' (my italics).[18]

Were anything like this to come to pass ever, the debate on tactical aviation would surely be foreclosed. It would be palpably absurd to invest in flying machines, a single exotic one of which might cost as much to build as nearly a 1,000 light tanks. But the improbability that the F-15 and F-14 will even start to phase out for a good 10 years yet has been underlined by the 1984 decision to adapt many F-15s to more truly all-weather strike tasks. There are, too, other signs that no such generational jump in costs is looming. Instead, the long escalation since World War II may be peaking out. For one thing, the levelling off in the maximum forward speeds of tactical aircraft has been remarkably abrupt (see Chapter 4). For another, the progressive easing under way in both the price and the bulkiness of electronics is making a most positive impact. In some contemporary warplanes, the electronic sector has directly accounted for 30 per cent of total flyaway costs; and the 'oft-quoted rule of thumb of $1,000 per pound' of electronic equipment affords a fair indication of the associated weight burden.[19] Thanks to the sustained dynamic of the electronic revolution, however, the percentage just cited is unlikely to rise further, while the weight penalties are set to diminish. Further capital economies are in train because of the extending use of composites and superior alloys, a lowering of specific fuel consumption and so on.

In fact, the F-16 has closely followed the F-14 Tomcat into large-scale production on the basis of a flyaway cost of $4.8 million (1975) in the USA and $6 million as manufactured by a West European consortium. Associated with this downward dip in the price trend is a 40 per cent decrease in all-up weight as compared with the Tomcat.

Yet here is another contrast which should not be read too glibly. The F-16 is configured more specifically for dog-fighting, the latter more for Combat Air Patrol well out. Also the F-14 is carrier-borne. True, the ubiquity and dexterity of the now veteran Phantom has shown convincingly how design for basing afloat now constraints performance in the air far less than it did in World War II. Nevertheless, the effect is still there. As much can be seen in the F-18 Hornet, the plane that has been replacing the A-7 Corsairs and F-4 Phantoms in USN carriers since 1982. A variant design known as the F-18L, comprehensively configured for land-based use alone, saved 1,375 kg of all-up weight by discarding some special equipment while economising on the undercarriage and other structural aspects. Moreover, it could save as much again through the consequent fuel economy. All-up weight would thereby be reduced by roughly a seventh though performance could be appreciably enhanced. F-18As, partially modified to fly economically just from the land, have been ordered by Australia, Canada and Spain.

What is more, the F-18 as such is of general import in this context. For while the F-14 left an exaggerated impression of cost-escalation in this field, the F-16 experience may have encouraged over-correction. What the F-18 signifies is a more gradual retardation of the upward curve. Thanks to its designers using more low-weight materials, its maximum weight is around 21,300 kg against 31,000 kg for the F-14. Correspondingly, its estimated unit cost has been kept down to the tolerably low figure of $26,000,000 on a production run of 1366.[20] That would be about $16,000,000 at 1975 prices.

Full-life Cost

What counts more in the final analysis, however, is the full-life expenditure on a system. In other words, the cost of operation year by year has also to be embraced. However, this dimension as yet enters the public debate less than does the original cost of development and procurement. Therefore, data tends to be more patchy if not more confused.

Here again, however, it is not too difficult to derive a sense of the proportions. A 1976 study indicated that to keep a squadron of B-52 or B-1 strategic bombers constantly on a 60 per cent ground alert would incur operating costs (at the price levels then current) of $50 million a year. The corresponding figure for FB-111 light strategic bombers or KC-135 tankers would be a little over $30 million. If the alert level were lowered to 40 per cent, the figure would be cut by a fifth in every case.[21] All the USAF squadrons in question operate 15 aircraft.

Some idea of how such sums are compiled has been afforded within the domain of tactical aviation. A 1977 reckoning for the annual operation, in peacetime, of a squadron of 24 F-16s was as follows — wages, $5,400,000; fuel, $2,100,000; spares and munitions, $2,100,000; and other items, $6,800,000. Thus the grand total was $16,400,000 as against $22,900,000 for a similar force of Phantoms.[22] In other words, each type of aircraft would cost three or four times as much to operate for, say, 15 years as what its own flyaway price had been.

When such calculations are projected forward, the biggest uncertainty becomes the future price of aviation fuel. For one thing, it is hard to gauge how 'elastic' this may be in relation to a shifting balance between supply and demand. For another, that balance is itself hard to predict, not least in respect of whether novel forms of fuel may eventually prove viable. The *Global 2000* report to President Carter concluded that, taking a secular or long-term perspective, petroleum production was still lagging behind demand. Moreover, the former would peak out before the turn of the century.[23] Conversely, Peter Odell, the Professor of Economic Geography at Erasmus University in Rotterdam, is one well-known authority who has consistently been optimistic on this score. Barely a year after the 1973–4 escalation in prices, he was suggesting that 'the longer term outlook for the build-up of significant oil production potential . . . is probably brighter now than it has been at any time in the whole post-1945 period'. Therefore, it 'will be possible for a conventional energy economy, based largely on oil and natural gas, to see the world's development through at least until the second quarter of the twenty-first century'.[24]

From divergencies like these stems an order of magnitude range in the predicted real price of petroleum at the turn of the century. Some analysts expect a fall of up to half. But others anticipate a rise of perhaps fivefold. Suffice here to remark that sanguiness as great as Professor Odell so regularly envinces is hardly warranted. In the study just cited, he further foresaw that with the 1980s would come 'market openings for Soviet oil in many parts of the world' (p. 64); and that OPEC would not prove 'maintainable' through 1980 (p. 227). The first judgement was wildly wrong; and the second too categoric.

So, conservatively, a price rise by a factor of 2.5 should be allowed for aviation spirit through the year 2000. After all, any expansion (through improved refining) of the fraction it represents of crude petroleum (seven per cent at present) is liable to be offset by steeply rising demands from civil aviation. Such an increase in the fuel bill of an F-16 force would raise its aggregate operating costs by a not inconsiderable 20 per cent.

In other respects, however, a broad tendency for the real costs of operation to level off or even diminish can be discerned. Witness calculations, in 1975 terms derived from the USAF Office of Air Force History study quoted above. These give $271 as the basic maintenance cost per flying hour for the F-86F in 1953, and $736 for the F-104A in 1958. With the F-4E, which entered service in 1967, the estimate rose only modestly to $883; and, judging from the data just cited, a corresponding figure for the F-16 might indicate a reversal.[25] As early as 1964, indeed, the one registered with the smaller and less exotic F-5A was $329.

A cardinal reason for the overall improvement has been a progressive reduction in the Mean Time Between Failures (MTBF) of key electronic subsystems. True, a vigorous debate erupted in the USA in 1980 over shortfalls on front-line squadrons in Europe, vis-à-vis the NATO stipulation of a 70 per cent level of aircraft availability. With the F-15, in fact, the proportion unavailable had too regularly exceeded a half.[26] Still, such aberrations need not affect the longer perspective too drastically.

Harder to ascertain from published evidence are the economics of carrier-based operation. The best approach is to compare the number of men involved with each aircraft with the 100 or so enlisted for every warplane in a Western land-based air arm. The *Forrestal*, which at 79,000 tons is quite representative of modern USN Carrier Vessels Attack (CVA), has a ship's company of 60 men for every aircraft on board. But that 'manpower slice' has to be multiplied several times to allow for all the other vessels in a task force, for rotational duty and for ashore support. So even apart from the lower sortie rates usually achieved at sea, this author doubts whether he should alter a broad inference he drew in 1962. This was that 'a carrier-based strike aircraft is, if all the respective supporting elements are weighed in the balance, several times more expensive to procure and maintain than a long-range heavy bomber'.[27] But a caveat to add is that carriers in general, and small carriers in particular, stand to gain rather more than shore bases do from the aforesaid easement of the maintenance task.

Though there is little indication as yet of the economic burdens recurrently incurred by the deployment of cruise or ballistic missiles in a ground-based mode, it would be unwise to discount them. Let us again take manning as a criterion. A Pershing 1 battalion has just 4 launchers and 54 missiles yet still 600 men. As regards the more traditional side of the artillery arm, evaluation is hamstrung by the way the organisation of guns into regiments and batteries varies. Perhaps a fair depiction of modern Western practice would be a field artillery regiment with an establishment of 18 towed guns and 360 men, a third of whom might actually be manning that weaponry. But to gauge properly the full 'slice' per gun, it is best to double this manning estimate — that is, to 40 men for every gun as opposed to the 100 or so just quoted for a tactical warplane. In other words, the manning differential between the two weapons genre is much less than that for capital outlay. Even with drones and RPVs, indeed, a similar narrowing is discernible. Thus the Aquila platoon being established in each US division will employ some 65 men to operate 20 RPVs.

But here a word of caution has to be entered against reading too much into inter-service comparisons of manning slices. Operating costs reckoned per soldier in uniform tend to be a lot less than with each member of an air or, indeed, naval arm. Never mind the fresh surge into tracked mechanisation (above all, self-propelled artillery) effected by so many armies these last 20 or 30 years. What British experience suggests, in fact, is that the differential has widened as aircraft have become more sophisicated. Dissection of the 1985/6 budget shows direct operating costs of $60,000 a year for every uniformed member of the RAF's general purpose forces, as against £27,500 *vis-à-vis* European theatre ground forces.[28] The percentage lead of over 100 thus obtained by the airmen contrasts with the 45 very similarly derived apropos 1967/8.[29] The implications will not be qualified much by the artillerists continuing to take more in the way of capital assets and highly skilled manpower than do, say, the infanteers.

Meanwhile, even more patchy evidence from the Warsaw Pact does lend support to the view that the economics of workaday operation militate less strongly against manned aircraft than do those of procurement. A Soviet field artillery regiment might typically have 1,000 men with 54 122 mm or 155 mm towed howitzers — again about 20 men per gun. In a battalion of BM-21 multiple rocket-launchers (each with 40 122 mm rockets and mounted on a lorry), there are 250 men to operate 24 of the weapons. What has to be remembered, too, is that the ratio of men to planes in Warsaw Pact air forces is still only 60 to one.

Economy through Multi-role?

Clearly, it would be quite wrong to read into the recently insistent rise in capital costs, in particular, proof conclusive that tactical aviation is inexorably pricing itself out of the battlefield environment, regardless of other relevant factors. The prognosis is not and never could be as trite as that. Measured in real terms, the said rise is now retarding. Besides which, the reckoning in of routine operating costs shifts the balance of advantage some way back towards aircraft as compared with some, at least, of the competitor systems. This being so, a general judgement about the future status of military aircraft must wait upon a synoptic appreciation: one that is broad and dynamic, not narrow and rigid. Moreover, its economic dimension ought to cover not only costs in peace but costs and advantages in war.

In the meantime, this much must be said. The financial strains air forces impose have already become sufficiently acute to encourage a search for radical remedies. Here is one reason why, through the 1970s, interest burgeoned in making aircraft more thoroughly 'multi-role'. This meant, most particularly, well suited to various modes of both bombing and air combat. It is redolent of how, in World War II, certain light bombers were adapted as long-range fighters. Britain's Mosquito and Germany's Junkers 88 were important examples.

Moreover, doctrinal traditions favourable to this approach can be traced back through that global conflict to the 'homogeneous' air fleets envisaged by Lanchester and the 'battle-planes' of Douhet. More immediately contributory, however, has been the diverse evolution of technology. The relevant gains include variable-sweep wings; more powerful engines; fuel economy; digital electronics; 'smart' ordnance; modular construction and detachable pods; and a revival of the cannon. The result should be that more versatility can be built into both the basic design of an aircraft and the specialist variants thereof.

Even so, the multi-role solution still involves compromises. The Australians have done well to stress the general point that the training load imposed on aircrew threatens to become a critical constraint.[30] There will also be technical objections that to conduct EW through ventral pods rather than via the best sites inside an airframe may be to compromise antenna performance. Perhaps a desire to avoid this within the wide arenas so common in naval warfare is why the USN is less keen on podding than could otherwise have been expected.[31]

Likewise, it is hard to adapt a plane to several roles without making it reflect too large a radar image. Podding can have this effect. So can internal overloading of the airframe. Thus the virtuosity of the Tornado as a heavy fighter or strike aircraft has meant that 'its broad radar cone lies in front of a large square-section cockpit flanked by two raked box-shaped intakes . . . Each tailplane is a simple large slab.'[32] Nor should one forget that, like all those World War II adapted bombers, the Tornado F2 interceptor has a relatively poor climb-to-height performance — still a drawback in respect of intruder tracking and inspection. Like them, too, it lacks the agility for close dogfights, despite the singular dexterity that comes with variable sweep. At 300 knots, for example, it can never execute more than a 5 g turn.[33]

Simple is Effective?

Already most, if not all, the advanced air forces have what is colloquially termed a 'high-low mix': a limited number of large and exotic planes for the more intensive contests, together with smaller and simpler ones for the less exacting or more subtle. So the question is whether the lowest layer in such a mix might not sometimes be pushed further down the sliding scales of size, elaboration and cost. Was Richard Garwin right to suggest that 'an equal-cost force of MiGs would just eat up the F-15'?[34] In fact, he was working to the dubious trade-off of capital cost of 7-to-1 in favour of the MiG-21. Regardless of that, however, could the essential point be valid, at least in dogfights? Is this a route to true economy?

Alas, the indications are by no means all affirmative. In the aftermath of its Korean experience, the USAF sought a lightweight 'air superiority' fighter. The successful bid, made by Lockheed, was to give rise to the F-104A Starfighter, a plane with a loaded weight of 13,250 kg and a flyaway cost of $3 million (1975).[35] Its chequered career was to include early withdrawal (mainly because of a lack of any scope for design 'stretch') from four squadrons of the US Air Defense Command. More notorious has been the loss in crashes of a third of the 750 F-104Gs the Luftwaffe procured. The Starfighter is a classic admonition against a specification being overloaded, be this literally or metaphorically.

More auspicious signs were later to be afforded by Northrop's F-5A Freedom Fighter: a quest for tactical economy in pursuit of high sortie-generation. It was designed specifically for the smaller and less sophisticated of the allied air arms, in Europe and elsewhere. The all-up weight of the derivative F-5E Tiger 2 is 11,000 kg which compares with 31,000 kg in the case of, for example, the F-15C Eagle. A Northrop study concluded that, while an F-15 costs four times what an F-5 does, the former averages one sortie a day versus 2.5 for the later.[36]

Granted, the F-5E has a burst speed of only Mach 1.65 whereas the F-15 can proceed a while at Mach 2.5. But how valuable is this margin? In 100,000 sorties by Mach 2.0 aircraft over North Vietnam, no combat time was recorded above Mach 1.8 and 'remarkably little' above 1.2.[37] Nevertheless, the better ratio of engine power to aircraft weight of the F-15 must still be useful in climbs and quick accelerations. In any case, the complete and successful reliance on stand-off missile-firing by Israel's F-15s operating against Syrian aircraft over the Lebanon in 1982 lends some credibility to the McDonnell-Douglas claim that 'In combat exercises against many types of aircraft (all having combat ability equal to or greater than an F-5), the F-15 exchange ratio was 88 to one'.[38] Israel's electronic domination shaped the course of that campaign and the outcome indicated.

Nor is a sense that while the Northrop approach is salutary it may also be too blinkered any the less firm now that this company has built a Mach 2.0 derivative from the F-5E. This is the F-20 Tigershark, which weighs 12,000 kg fully loaded. Never mind Northrop's insistence that, even though this machine could generate 40 per cent more sorties than does a General Dynamics F-16, it needs only half as much support manpower.[39] In June 1985, General Dynamics offered to build a slimmed-down version of the F-16C which could be superior to the F-20 in every respect except sortie frequency and instantaneous turn rate. In particular, it would retain the F-16's much superior bomb load. It is claimed that, on a production run of 216 machines, the flyaway cost of the former would be 20 per cent less than the latter's. It would also have a lower life-cycle cost.

Questions are inevitably raised by anything that smacks of a wilful retreat from sophistication. For this may involve a casting aside of our qualitative edge at the very time when the Soviet air force is earnestly seeking qualitative improvement. Nevertheless, these last 15 years or so have seen an accent being placed everywhere on the light strike-trainer. Since its primary task is pilot-training, it has to be two-seated. Even so, it will be decidedly non-exotic. By 1983, two of the West's exemplars — the Alpha Jet made by Dassault-Brequet/Dornier and the Hawk made by British Aerospace — had been ordered world-wide to the tune of 480 and 260 respectively.[40] According to one survey, the non-Communist world now has to have a 'trainer population of around 8,000 — 4,000 propeller-driven aircraft for primary and basic training and 4,000 jets for basic and advanced tuition'?[41] But as with the old Harvard propeller-driven trainers used in Vietnam, such machines are inherently eligible for certain combat tasks. Although these are mainly strike, seventy-five British Hawks have been given a pair of Sidewinders to contribute to air defence.

All the same, those tasks may be too circumscribed in relation to the price that still has to be paid in material and human resources. Data available about the export sales of the Alpha Jet and the Hawk bespeak a flyaway cost for each of *c*. $4,500,000 at current prices. By various indices, however, their capacities are well below those of, to draw again the most apposite comparison, the F-16. Most Hawks lack air-to-air armament while the Alpha Jet can only carry, on a centre-line pod, a 30 mm cannon. The initial rate of climb of the Alpha Jet is given as 57 metres a second as against the 312 of the F-16 or, for that matter, the 160 of the F-5E.

Nor may strike trainers be all weather in any very complete sense.

Nor may they be adapted to diversified EW, notwithstanding a place for a second crew-member. Nor can they bear long-range missiles. Nor are variable-sweep wings an available option. For even in much larger aircraft, the outboard pivots required in that technology contribute 3 or 4 per cent of the empty weight,[42] a fraction that must be multiplied several times to allow for the 'knock-on' effect (see Chapter 4). It does seem, in fact, that a loaded weight of 17,500 kg (as per the Sukhoi-17) is close to today's lower limit of practicality for variable sweep. Nor is this limit likely to come down very fast or far.

All in all, the signs are that strike trainers would prove as disappointing in a really dynamic war as Israel's Magister force did in 1967 (see Chapter 1). In other words, Vietnam was atypical. Even so, a much remarked paper submitted to Britain's Ministry of Defence in 1979 sought to push the advocacy further. It called for the development and mass-production of a Single-Seat Propeller-driven Fighter (SSPF) with an all-up weight of only 2,500 kg. Forty per cent of that modest mass might comprise 100 of the Wasp anti-tank missiles then under development, each being intended to seek out an individual target up to 10 km away.[43] Alternatively, fewer Wasps might be combined with a brace of Sidewinders or machine guns or 20 mm cannon.

An air-cooled piston engine (rugged and with a diffused thermal image) would, so it was claimed, be adequately immune to light anti-aircraft fire. Further immunity might be conferred by an ability to approach targets at a speed of 400 km/h yet a height above the surface of but 12–15 m. Putatively, too, its base facilities could be two orders of magnitude simpler than is the case with more elaborated planes.

Assertions that such an aircraft could be mass-produced for $200,000 or thereabouts aroused immediate scepticism. More importantly, notions that the tactic of 'ultra-low' infiltration might keep the loss rate in all-out war as low as, say, 40 per cent a week could only be dismissed outright.[44] Still, the tactical concept in question could hold out more promise if it were to rest on the full exploitation of the self-homing propensities of weapons like the Wasp by using them for indirect fire from concealed locations (see Chapter 13). Never mind that the Wasp itself has since been curtailed.

A dispute which is similar though more esoteric is whether the E-2C version of the USN's carrier-borne Hawkeye should not have been preferred to the much bigger E-3A Sentry when NATO came to select an Airborne Early Warning and Control System (AWACS). Ostensibly at least, the Hawkeye can maintain a flow of track data (from sensors active and passive, on-board and external) comparable with that from the E-3A. A typical claim for the former has been that, in the active radar/IFF mode, up to 300 air or sea 'targets' can be tracked simultaneously, this out to a distance of 325 km. Even on a passive basis, up to 256 such targets might be tracked.[45] As many as 40 actual engagements (air-to-air and air-to-surface) may be directed concurrently.[46] What is more, there is considerable scope for progressively improving the speed and memory of the E-2C's computer.

True, the Hawkeye's smaller radar may be the more susceptible to side-lobe jamming. On the other hand, its radar frequencies are well attuned to the scanning of choppy seas. Initially at least, the E-3A was deficient in that domain.

Where a difference between the two models is most evident is in the human contribution. The Hawkeye carries an airborne tactical data team of three, whereas that in the Sentry is 13 strong. Maybe the former arrangement imposes too heavy a workload. Alternatively, the latter may assume more of the actual command function than may be appropriate to an airborne platform. For it may be even more susceptible to EW interference or outright destruction than corresponding facilities on hand. But at least the E-3A can attain a speed (in level flight) of 1,000 km/h whereas the E-2C manages a mere 560 km/h. Even so, a Sentry on patrol in central Europe would normally stay 275 km behind the FEBA, the better to evade enemy intrusions.[47]

Then again, the Hawkeye can stay but four hours on a patrol station only 370 km from base, whereas the E-3A can loiter for seven at 1,850 km. So perhaps the most important question is not which plane is most capable electronically. Instead, the nub may be that the E-2C is not free-ranging enough to cope with wide fronts or avoid fast-moving predators. Nevertheless, it has been acquired by both Israel and Japan; and has been seriously considered by France. Nor must we overlook the moves the USAF made to load the evaluation procedure in favour of the Sentry.[48]

Undoubtedly, too, the trend is towards AWACS machines with an all-up weight close to the Hawkeye's (24,000 kg) rather than the Sentry's (150,000 kg) or its even bigger Soviet counterpart, a variant of the Ilyushin-76. But this is another aspect of air warfare in which it would be wrong to push too fast and too hard the quest for simplicity and compactness, at least when the intention is to use the finished product against strong and sophisticated opposition.

References

1. W. D. White, *U.S. Tactical Air Power* (Brookings Institution, Washington DC., 1974), fig. 4.2.
2. E. S. Woytinsky, *A Profile of the US Economy* (Praeger, New York, 1967), table 15.5.
3. *Defense and Foreign Affairs Digest*, 11/1978, p. 26.
4. White, *U.S. Tactical Air Power*, pp. 46–7.
5. S. J. Dudzinsky and J. Digby, R-1957-ACDA, *Qualitative Constraints on Conventional Armaments: An Emerging Issue* (Rand Corporation, Santa Monica, 1976), p. 88.
6. 'B-1/Stealth Competition Emerges', *Aviation Week and Space Technology*, vol. 120, no. 9, 27 February 1984, pp. 16–18.
7. *International Defense Review*, vol. 9, no. 2, April (1976), p. 167.
8. Desmond Ball (ed.), *The Future of Tactical Airpower in the Defence of Australia* (Australian National University, Canberra, 1977), p. 64.
9. *Guardian*, 20 January 1978.
10. Dudzinsky and Digby, *Qualitative Constraints*, ch. 3.
11. *Defense and Foreign Affairs Digest*.
12. Paper presented to an International Seminar at Erice-Trapani in Sicily in August (1981).
13. Desmond Ball, 'The Costs of the Cruise Missile', *Survival*, vol. 20, no. 6, November/December (1978), pp. 242–7.
14. *Aviation Week and Space Technology*, vol. 103, no. 15, 6 October 1975, p. 20.
15. R. L. Pfaltzgraff and J. K. Davis, *The Cruise Missile: Bargaining Chip or Defense Bargain?* (The Institute for Foreign Policy Analysis, Cambridge, 1977), p. 15.
16. 'Simple Army Drone Grows Complicated, Expensive and Late', *Wall Street Journal*, 23 November 1984.
17. *Military Technology*, 10/83, October (1983), p. 24.
18. White, *U.S. Tactical Air Power*, p. 54.
19. A. D. Slay, 'An Air Force Avionics Policy', *Air Force Magazine*, vol. 60, no. 7, July (1977), pp. 30–9.
20. *Aviation Week and Space Technology*, vol. 115, no. 1, 6 July 1981, p. 24.
21. A. H. Quanbeck and A. L. Wood, *Modernising the Strategic Bomber Force* (The Brookings Institution, Washington DC, 1976), Appendix A.
22. *Aviation Week and Space Technology*, vol. 106, no. 18, 2 May 1977, p. 15.
23. *The Global 2000 Report to the US President* (Penguin, Harmondsworth, 1982), vol. 1, p. 27.
24. Peter Odell, *Oil and World Power* (Penguin, Harmondsworth, 1975), p. 226.
25. Desmond Ball (ed.), *The Future of Tactical Airpower*, p. 87.
26. *Flight International*, vol. 118, no. 3716, 26 July 1980, p. 275.
27. Neville Brown, *Strategic Mobility* (Chatto and Windus, London, 1963), p. 119.
28. Cmnd. 9430-11, *Statement of the Defence Estimates, 1985* (Her Majesty's Stationery Office, London, 1985) tables 2.3 and 4.1.

29. Neville Brown, *Arms Without Empire* (Penguin Books, Harmondsworth, 1967), p. 131.
30. *Australian Defence* (Australian Government Publishing House, Canberra, 1976), p. 24.
31. 'Pods: That Something Extra', *Aviation and Marine International*, 2, 1981, pp. A1 to A8.
32. *Flight International*, vol. 114, no. 3632, 25 October 1978, p. 1585.
33. *Flight International*, vol. 125, no. 3907, 24 March 1984, p. 764.
34. Dudzinsky and Digby, *Qualitative Constraints*, ch. 2.
35. M. S. Knack, *Encyclopedia of US Air Force Aircraft and Missile Systems* (Office of Air Force History, Washington DC, 1978), vol. 1, pp. 175–90.
36. J. Fallows, 'America's High-Tech Weaponry', *Atlantic Monthly*, vol. 247, no. 5, May (1981), pp. 21–33.
37. Pierre Sprey, 'Mach 2, Reality or Myth', *International Defense Review*, vol. 13, no. 8 (1980), pp. 1209–12.
38. Fallows, 'America's High-Tech Weaponry', p. 24.
39. *Defense and Foreign Affairs Digest*, vol. XI, no. 132, 2 July 1982, p. 1.
40. *Flight International*, 14 April 1984, vol. 125, no. 3910, p. 999.
41. Graham Warwick, 'The Combat Trainer Market', *Flight International*, vol. 120, no. 3755, 4 July 1981, pp. 24–5.
42. 'Variable Sweep', *Flying Review International*, vol. 24, no. 1, September (1968), pp. 70–1.
43. Warwick Collins, 'Mustang with a Kick to Stop Russian Tanks', *Daily Telegraph*, 27 September 1980.
44. See letter from Warwick Collins in *Daily Telegraph*, 15 October 1980.
45. *Flight International*, vol. 119, no. 3762, 13 June 1981, p. 1843.
46. *Aviation Week and Space Technology*, vol. 115, no. 9, 21 August 1981, p. 47.
47. *Flight International*, vol. 119, no. 3757, 9 May 1981, p. 1295.
48. David Harvey, 'Fire in the Sky: the Airborne Early Warning debate', *Defense and Foreign Affairs Digest*, 10/1977, pp. 10–19.

13 A MULTIPLE REVOLUTION IN WAR

Air Power Ascendant

In World War II, overland offensives rarely succeeded except against a background of aerial supremacy. Moreover, in most cases where this did not obtain, poor weather prevented the full exercise of air power by either side. The Soviet envelopment of Stalingrad is the most conspicuous example.

Then for a good 30 years, aviation was to be ever more ascendant in war. True, Israel's supremacy in that sphere was not made manifest as sensationally in 1973 as six years' earlier. Even so, it again ensured her enough of a victory without an unbearable toll in Israeli lives. Much the same applied in the Lebanon in 1982.

Nor is this perspective invalidated by the inability of United States air power to redeem South Vietnam. All else apart, things might have turned out differently had the precision-guided generation of air-delivered ordnance arrived before a collapse of resolve within the American public as a whole and the young in particular. Let us never forget that the final Communist push to Saigon was no spontaneous peasant revolt. It was an armoured blitzkrieg, 1940 style.

Today, high priority is accorded to air power by both the major alliances. Comparisons invite themselves with the classic era of the blitzkrieg, early in World War II. For this was a time when tactical aircraft served only too effectively to spearhead mechanised thrusts against troops mostly lacking, just as the great majority would be now, previous combat experience. Yet the weight of aviation deployed was low by modern standards. Admittedly, in the campaigns across the mountainous and maritime flanks of Europe (Scandinavia and the Balkans), the Germans did employ what would still be thought a large number of warplanes relative to ground forces. But in the massive overland offensives against the Low Countries and France in 1940 and then the USSR in 1941, the Luftwaffe had only 20 to 25 aeroplanes committed for every army division in the Wehrmacht order of battle. In central Europe today, the corresponding ratio is 65 for NATO and 50 for the Warsaw Pact.

Nor does that comparison tell anything like the whole tale. It takes no account of a huge expansion in the number of aircraft (meaning, most particularly, the introduction of many hundreds of helicopters) operated by the armies themselves. Nor does it of the much bigger part surface weapons today play in air defence. Nor can it in the ability of hundreds of modern warplanes to sortie over the central area from bases in Britain, France, Spain and, of course, the USSR. Also the prospect is that, during the first week or two of a warlike crisis, both sides would reinforce their air arms in the European theatre proportionately more than they would their land.[1]

Besides, the fact that aircraft costs have risen a lot faster since 1945 than have those of land weapons (see Chapter 12) does relate considerably to a differential growth in sheer size. In 1944, the RAF received into squadron service its first jet-propelled fighter, the Meteor. Its closest contemporary counterpart — a variant of the Phantom — has three times the loaded weight. When used in a strike role, the Phantom carries eight times the weight of bombs a Meteor ever could.

Outside Europe, contrasts tend to be less stark in the terms here indicated. As a rule, an aerial order of battle is much less easy to relate arithmetically to a land one. This is because of a whole range of variables, from quality of technology to local geography. Two things are safe to say, however. Since 1945, most of the air forces in Afro-Asia and Latin America have arisen completely or virtually from scratch, which is something that cannot be said of most armies. Also, there are certain sectors in which the weight of air power is demonstrably high, viewed again in historical perspective. Take the build-up that has occurred in the Soviet Far East, with the 7,000 mile border with China as its prime *raison d'être*. In this theatre, 52 Soviet line divisions are supported by 2,700 air-force warplanes: an indicated 52 per division, a figure which rises to 80 if one adds in a due proportion of PVO Strany — the strategic air defence command.[2] Clearly, such addition is well justified by the conceptual changes identified in Chapter 2.

The exceptional stress thus laid on theatre air power is partly a response by Moscow to a low ratio between ground troops and linear space across a front with weak road and rail links yet extending over mainly open terrain with a largely open climate. But it must owe something to the plain truth that, contrary to what has sometimes been supposed, the USSR's rearmament drive in the Far

East has been aimed more at intimidating China than at guarding Soviet borders against her.³

No matter, the crucial issue now is whether we will soon see a peaking out of the influence of 'air power' on war or threats of war. Ever since the term was coined, apparently by H. G. Wells in *War In The Air* in 1908, its core connotation has been 'that the third dimension above the Earth is actually exploited to advantage by the platform or vehicle' located up there. It is a perception which easily merges with the celebration of the pilot's role observed in Chapter 3. To which one might add that it also encourages air staffs to try and preserve their corporate identity through an insistence that the 'proper application of offensive air power is against targets that are beyond the reach or the capacity of other weapons systems'.⁴ But what does 'proper' mean in relation to the future? And what of the operational tasks today managed from the cockpit? Might some or all of these gradually become less viable? May viability be preserved only by delegating more and more of them to unmanned systems?

Finally, are the institutionalised distinctions between air, land and sea power losing whatever utility or relevance they may once have had? In other words, has F. W. Lanchester come right after all? He it was who dubbed military aviation 'the fourth arm': one which stood alongside the infantry, cavalry and artillery rather than over against the Army or Navy as such. From which he said it followed that 'The command of the air can never be taken to carry a meaning so wide or far-reaching, or in any sense as comprehensive, as that understood when we speak of command of the sea.'⁵

As one explores these interactive questions, one needs must steer a middle course between a skittish radicalism on the one hand and an atavistic or obfuscatory conservatism on the other. In this milieu, atavism most readily finds expression in single-service partisanship. But obfuscation may have different roots. Since 1945, there has been a strong drive, well manifest in staff training and in defence management, to break down interservice rivalries, especially as between air and naval arms. Alas, the result today, almost the world over, is an ambience in which the three recognised services may be too ready to support one another's pre-existing stake in the order of things!

250 *A Multiple Revolution in War*

Warplanes in Decline?

Still, skittishness is also more than amply in evidence. Yet again, the imminent demise of the tactical warplane is regularly being forecast, the grounds being the habitual ones that it is allegedly becoming either impossibly expensive or else hopelessly vulnerable. However, the former thrust of argument, epitomised by the Brookings study mentioned in the last chapter, is being undercut not just by a slow-down in the rise in flyaway costs but by other contemporary tendencies. Aircraft are becoming easier hence cheaper to maintain, not least in the field. Meanwhile, in relation to many of the offensive scenarios one can plausibly depict, strike planes should soon be one to two orders of magnitude more lethal than they were even 15 years ago. Improvements in ordnance and in on-board electronics are, of course, the main reasons for this. Besides which, many positive trends, from better engine performance to advanced digital computing, are conducive to aircraft being more effectively multirole, always provided their crews can cope with the extra training and combat stress such dexterity requires. The evolution of the F-15E as the USAF's enhanced Dual-Role Fighter exemplifies the point.[6]

Clearly, however, still more attention has also to be paid to raising sortie rates in war because this is among the most efficacious ways of making aircraft effective *en masse*. What may have to be conceded, too, is that keeping planes in squadron service until they are all but worn out may be less viable than in the past. Granted, two factors work the other way. Recourse to modular replacement (including the progressive updating of the ordnance borne) will make it easier to keep old airframes usable. So, too, should greater structural endurance. Yet these gains may avail military aviators little once the whole concept around which a given machine was originally designed has been overtaken by manifold changes in an operational ambience evolving at an unprecedented pace. Therefore, shorter cycles of replacement will be liable to negate in part any easement of the economic pressures in certain other respects.

Hopefully, however, the rate of obsolescence will usually be slowed by the synoptic anticipation of change: political, technological and tactical. Associated with which is the nagging question of how far one dare push the claim that 'Simple is Beautiful'. Can real savings be achieved by directing aeronautical science to the ends of economy and availability rather than have it preoccupied with exotic capability.

An instructive test case is that afforded by the F-20 Tigershark: the Northrop private venture that attracts much interest but no sales (see Chapter 12). So perhaps we should pursue further the comparison with the F-16 Fighting Falcon, a model half a generation earlier and just a third as large again. With a 'fly-away cost' closely equivalent to that for the F-16 but normal operating costs put at 50 per cent lower, the Tigershark could be expected to cost 40 per cent less across, say, a 15-year life. Yet the higher operational sortie rate said to go with this (Chapters 1 and 12) does imply, other things being equal and allowing for the Lanchester square law, a doubling of effectiveness, force for force.

In air-to-air combat at medium altitudes or above, however, 'other things' will not be at all equal because the Tigershark will be too deficient in engine thrust and in respect of the weaponry it bears. With Close Air Support and shallow interdiction, on the other hand, these shortfalls in thrust and warload might sometimes matter less. The same goes, too, for air interception at low altitude. What may be a critical factor is the electronic environment. At all events, the F-20 will continue to fail to attract orders because one has crashed and because it is (a) out of phase with crucial re-equipment cycles; (b) not multirole enough; and (c) too obvious a Northrop foray against alleged Pentagon obfuscation. Even so, it does exemplify the scope afforded by new materials and technologies.

So casting the argument more widely, the 'bottom line' must surely be this. Air forces should never equip themselves with manned aircraft so rudimentary in concept that they are liable in consequence to suffer much higher rates of loss. Think of Argentina's loss in action of 25 of the 30 Pucara light strike planes deployed to the Falklands: 6 destroyed in the air plus 19 on the ground. This was in exchange for one British light helicopter shot down and one warship slightly damaged. The connotations for morale in battle are obvious. Hardly less worrisome, however, may be an insufficiency of aircrew recruits prior to hostilities. Even the commander of the Israeli air arm has lately expressed concern about a pilot shortage, as things stand.[7]

Not that sharp rises in representative rates of attrition are likely to be avoided, whatever planes are used. As intimated already, the vulnerability of aircraft in flight seems not to have got much worse these last two to four decades. Almost certainly, it has done so less than that of tanks in close-order attack or artillery in emplacements or frigates in contested waters. None the less, percentage losses do seem poised to increase, especially with reference to incursions into hostile air space. At the higher altitudes of approach, of course, this shift in advantage is already well advanced. These days manned aircraft could rarely survive several miles up the way that, in spite of everything, strategic bombers did in World War II.

All the same, the actual prospects at all altitudes will always be shaped not only by the mean trend towards more lethal encounter but by a host of contingent circumstances. Geography will be among them. So will the quantitative and qualitative balance of forces, not least in their electronic dimension.

Important in this sphere will be the shifting balance between warplanes and SAMs. Two considerations may particularly affect it. The one is that, according to Aérospatiale (a firm experienced in this matter), the application of ram-jets to tactical missiles received a distinct fillip around 1975 in that it proved easier than before to move from the rocket-booster phase to the sustained high speed (e.g., Mach 3.5) that ram-jet propulsion facilitates. The other is that, given the aerodynamic difficulties inherent in lateral manoeuvre in rarified air, much scope remains for improving the agility of manned warplanes at high altitude before running up against the limits of aircrew tolerance of centrifugal stress. Nearer the surface, on the other hand, that ragged but decisive threshold is now hard upon us, to the detriment of the warplane whether as interceptor or intruder.

Meanwhile, the neutralisation of anti-aircraft defences (by electronic jamming or deception and by actual attacks on airfields, master radars and artillery sites) promises to figure prominently in any air offensives conducted in the near future. Sometimes endeavours to suppress such resistance will be integral to attacks on other targets, including those which present themselves at sea and during Close Air Support. In the CAS milieu, field artillery and other ground elements will be called on to assist this purpose.

What NATO's 1976 doctrinal statement implicitly accepted, however, was that the classic quest for untrammelled mastery of the skies is no longer a valid aspiration, regardless of the tactics employed. Behind this acceptance lay an awareness of the sheer density with which surface-to-air weaponry may be deployed nowadays, especially in the forward areas. Take, for example, a US

A Multiple Revolution in War 253

Army analysis, likewise published in 1976, of the Soviet panoply in this regard. From it, one can readily infer that there will be some 125 crew-served anti-aircraft weapons in a division. Over 80 per cent of them will be deployed within 40 km of the Forward Line of Enemy Defence (FLED).[8] Here alone is a daunting suppressive task. Yet these crewed systems will be complemented by a 'dense blanket' of machine guns, hand-held SAM-7s, SAM-9s and SAM-13s: the last two being SAM-7 derivatives which are mounted on tripods on wheeled and tracked chassis.[9] Altogether, of course, several scores of thousands of special anti-aircraft pieces are deployed within the Warsaw Pact.

Still, one mode of suppression that cannot be negated by the sheer density of the defence is the broadcasting of chaff. Another is the use of Remotely Piloted Vehicles (RPVs) to act as decoys: that is to say, to project images that, on radar screens at any rate, simulate well approaching warplanes. What is more, the latter is an aspect of Electronic Warfare (EW) in which the West is even now making rapid strides, whereas the USSR is not: a contrast no doubt related to miniaturisation being the *sine qua non* of such devices.

However, neither RPVs nor chaff are very effective at the short slant ranges that are presented during low-altitude intrusion. Moreover, it has to be said that, them apart, every technical trend is inimical to the suppression from whatever altitude of surface-to-air defence. Anti-aircraft weapons show every indication of becoming progressively more compact yet capable, this for some time ahead. They should also become less reliant on external co-ordination, which means *inter alia* that it will be harder to cripple defences by wrecking master radars or command posts.

Collaterally, the tendency for systems to be dual-purpose (meaning, above all, to be usable against both *Tank und Flieger*) is sure to strengthen. The facility with which Vertical Launch Systems may fire off a whole variety of ordnance will be conducive to this. So may be a disinclination on the part of those who design battle tanks any longer to armour them so solidly. In other words, a conviction could burgeon that, when tracks can be blown off by scatterable minelets or whole vehicles enveloped in Fuel Air Explosive (FAE) regardless of steel cladding, there is little to be gained in persevering with mechanical leviathans intended to crash through the opposition. Therefore, on those (perhaps infrequent) occasions when the armoured fighting vehicles of the future do present themselves over open sights, they may then be engaged by, say,

40 mm cannon shells fired at a high rate. Yet such a response could also be highly effective against planes. Naturally, the spread of this kind of ambidexterity would itself make denser still the opposition faced by attacking aircraft.

Moreover, the responsiveness of individual surface-to-air weapons looks set for critical improvement. Data adduced in Chapter 7 about representative reaction and pursuit times suggested that marginal savings in either would often lead to a substantial rise in the success rate of 'line of sight' interception. In essence, this is because manned aircraft have little scope left for shortening their times of transit over their target areas, assuming the men on board are to remain actively involved in the tactical management of attack and evasion. Of particular relevance is the greater dexterity and autonomy of missiles which is connoted by the concept of Vertical Launch and that of Self-Initiating Anti-Aircraft Missiles. For this extra facility should often allow the pursuit of an intruder well beyond the line of sight observable at the launch pad. Add to this the continuing development of 'look-down, shoot-down'. Then it is hard not to conclude that aircraft flying low through well-defended air space will come to have poorer chances of survival. Over the next 20 years, indeed, this trend probably will be steeper than the corresponding ones for systems deployed on the surface of the land or the sea.

Needless to say, this conclusion is underlined by the truism that the short distances and brief contact times presented in the surface-to-air engagement of low-flying aircraft do not leave much scope for improved screening by means of decoys and chaff. Nor can these stratagems be complemented at all decisively, at that level, by stand-off jamming, Stealth technology, evasive manoeuvre or still tighter contour-hugging. From which the inference might be drawn that reversion to higher altitudes will normally be the best answer for the execution of strike missions. But few analysts or operators would actually subscribe to that. The geometrical attractions of low approach are too compelling, certainly in central Europe.

However, the one big caveat remains in respect of all aspects of anti-aircraft defence. This is that efficiency is liable to be reduced a lot (above all, at the outset) by the interaction of the various stress factors: surprise, shock, fear, fatigue, darkness, bad weather, Electronic Warfare, lack of real war experience and so on. After all, EW and bad weather may all too easily combine to reduce the range of detection by radar well below the notional limits of a particular type of set presented with a certain kind of aerial target flying at a given height in a particular landscape setting. What is more, the image actually projected onto a screen may be interpreted with less than perfect aplomb by a human operator in an advanced state of confusion.

The Sudden Blitzkrieg

As is at last coming to be appreciated, the horrendous stress and attrition surprise attack can produce demands far more attention than, let us say, the solitary paragraph the subject received in Liddell Hart's *Deterrent or Defence* (see my Chapter 5). Even in terms of the trite arithmetic of targeting, the dividends that accrue from getting the first sortie and salvo in are much augmented by the lethality of modern weapons, not least air-delivered ones. Over and above which, the psychological shock of dynamic war is bound to be severe as those first blows strike home. Speaking in the spring of 1944 about the forthcoming 'Second Front', Field Marshal Rommel said that D-Day itself would be 'the longest day'. In a war in central Europe in 1995, it might be more a question of the longest hour.

Among the applications of tactical air power in the face of a sudden blitzkrieg is the monitoring from aerospace of a potential aggressor. Sometimes this might be the means of ensuring that such elementary precautions as the deployment of troops to battle positions are taken in good time. Even so, no amount of monitoring is going to negate at all completely the impact, psychological no less than material, of an enemy first strike. Suppose one even knew in advance the planned time, strength and frontage. The actual onslaught would still be the most dreadful of surprises.

To an extent, however, the initial shock of battle acts on both sides. Where it may particularly do so is in the proneness of surface-to-air defences to exhibit a 'learning curve': the interval required for efficiency in real war to rise near to what was normally attained in peacetime exercises. Undoubtedly, such a curve will impose itself as long as mere humans remain at all involved in operating weapons. On the other hand, simulator training together with electronic aids in battle may well compress the typical span from several days to a few hours, albeit dreadfully long ones!

But this prospect encourages, in its turn, the view that one of the biggest contributions aircraft can make lies in checking an enemy offensive at the outset or maybe, some may hint, aborting it before commencement. Among the relevant factors to bear in mind are what could prove strong political inhibitions against ordering friendly ground troops to ready themselves properly for war or, in the event of tension proving prolonged, holding them in that expectant state. Everything else apart, the convergence to be observed or anticipated, in land warfare, as between defensive postures and offensive ones will make some parties all the more concerned never to dispose their ground troops in a manner that might be deemed 'provocative' (see Chapter 5). Air arms have customarily been able to ready themselves far more discreetly. A major emphasis on VTOL coupled with dispersal in crisis could diminish that contrast. But it will not eliminate it.

So how well-founded is the claim that air operations in wartime (above all, those air-to-surface) correspond a lot more closely to formation training in peacetime than might apply with, say, a tank-and-infantry attack? In other words, are the learning curves less steep and prolonged? If so, this is ostensibly because of the *modus operandi* and also in respect of the psychic disconnection involved. Yet any such difference can be overdrawn, even when related simply to the demands made on aircrews in flight. Nor should anyone forget that, more or less as soon as fighting commenced, many airfields would cease to appear as rock-solid citadels (see Chapter 9) and become instead foci of contention.

Nevertheless, this much can be allowed. In several respects, aircraft are well suited to intervene massively at the outset of a land conflict. These are (a) the speed with which they may respond to an overground sally; (b) their awesome ability to concentrate to whatever extent is required, more or less regardless of terrain; (c) the feasibility of a lot of mission planning, and even rehearsal, well ahead of any hostilities; and (d) the way strike missions admit of single-minded concentration by a few highly motivated individuals. At this early stage, in fact, emotional stress may stimulate positive reactions. Eventually, however, it may vitiate them, especially if casualties are high.[10]

For this reason alone, it is far from certain how well air echelons might withstand the strain of a major conflict extending across days rather than hours. This strain would largely be caused by the progressive undermining by losses in action of morale, cohesion and sheer numerical strength. Still, it would also stem from other features of modern combat, including having to operate more or less 'round the clock' and in the face of chaff or other ECM or maybe poor light and adverse weather.

Nor must one wax too sanguine even about the initial riposte. Many aircraft earmarked for it might themselves be caught on the ground. The rest might have to take off into the dismal 'small hours'. This is because that time of day, not 'first light', may come to be identified as ideal for the unleashing of aggressive war. Likewise, such a demarche might also have been arranged to coincide with a spell of bad weather, thereby gaining for the advancing columns something of the immunity from air action thus initially secured by the Germans during their Ardennes offensive in 1944 or by the Chinese during their entry into Korea in 1950. Only a prolonged and widespread 'black-out' of meteorological data (against a background of political tension) would preclude quite reliable forecasting a week or two ahead. What is more, even the volatile climate of central Europe regularly manifests what meteorologists term 'persistence of type': a quite pronounced tendency for a given weather pattern (anticyclonic, static cyclonic, mobile cyclonic, etc.) to last several days.[11] Other places where the same applies include Korea and parts of the Middle East.

Agreed, electronics can make such factors more manageable, not least from the air. True, too, that dynamic war may also overstretch artillerists, tank crews, sailors and so on. However, the cockpit environment will always impose a singular blend of stress factors: physical constriction and isolation; the diversity and urgency of the tasks to be performed; and the rather ethereal relationship that may obtain, via the sundry electronic aids, with an ambient environment that may be drear and menacing in every way. Nor does cockpit redesign promise to ease at all decisively the grim interaction of work and anxiety a single crewman may experience. That much is acknowledged by the USAF's choice, in 1984, of the F-15E as its Dual-Role Fighter design. For this is twin-seat, whereas the standard F-15C (which focusses much more firmly on air superiority) bears only a pilot.

Among the first indications of fatigue in aircrew is what is known as 'coning'. With it, the single-mindedness alluded to above narrows to an obsession with the primary aim, flying to the objective or whatever. In this way, the adaptive dexterity extolled as the supreme virtue of the manned aircraft is atrophied. Accordingly, quite the hardest task in tactical air planning today is mapping out a profile of sortie frequency over time. It has to be one that can keep within tolerable bounds of cumulative fatigue and the sharp fluctuations in morale that aircrews are subject to in battle. But the ultimate test has to be the influence the aircraft then have on the course of the land war.

More fundamentally, it has become imperative constantly to review the whole structuring of air power in relation to the demands of dynamic war: meaning, first and foremost, the shock of surprise attack. Our prime concern ought surely to be to mobilise technological excellence to make more resilient the individual systems and the air panoply as a whole.

Resilience is an attribute to seek by diverse routes. Avoiding undue elaboration in aeronautical design is certainly one. Another may be to relieve manned aircraft almost entirely of responsibility *vis-à-vis* the delivery of nuclear weapons. Agreed, it is possible to envisage some scenarios (especially outside NATO Europe) in which an aircraft's unique ability to fly in a holding pattern with its nuclear bombs on board could contribute materially to crisis control. But this would mainly apply, assuming it did so at all, to initial nuclear release.

Next is the need to make air defence less prone to disruption through action against its command-and-control network. Here a two-fold approach may be required. The one would be to exploit the technical trend towards more capable aerial platforms by designing aircraft that can engage low-flying adversaries at quite close quarters. A defending plane or rotorcraft at, say, 1,500 ft might be well placed to shoot down from above an intruding aircraft or, indeed, cruise-missiles obliged to stay within 250 ft of the surface for fear of the surface-to-air weapons (see Chapter 8). What such a posture might depend on, however, is control arrangements that afforded full scope for local co-ordination between air and ground.

The other path whereby to avoid the disruption just apprehended is to make a theatre network of command and control less prone to crippling attack. Various answers commend themselves. They include making the ground-based modes of aerial command as mobile as possible, a remedy that NATO's 2nd Allied Tactical Air Force (in north central Europe) has actively explored since 1983. Here the concept is that scores of thousands of square miles of air space are controlled synoptically by a mobile Sector Operations Centre (SOC) set under 'an 18-ft dome of air-inflated material

A Multiple Revolution in War 259

arching between two rows of chassis-mounted cabins'. Inevitably, however, antennae exposure is a problem even though they may be set well away from the SOC itself. Besides which, to switch a SOC to a new location and make it operational can take up to 24 hours.[12]

Another precaution may be to make the AWACS echelons less exposed to disablement, be this in the air or on the ground. Take, for example, the AWACS coverage of NATO Europe and the maritime approaches thereto. The intention is to provide this through 1995 with 18 Sentrys in a NATO joint force plus eleven Nimrod Mk 3s in the RAF. Effectively speaking, the former only operate from one base, Geilenkirchen, so far as central Europe is concerned. Yet all such machines must be terribly susceptible to immobilisation on the ground as well as considerably so, in some situations, to interception in the air. More attention ought to be paid to whether this weakness might prudentially be reduced through a combination of extra procurement (despite the high unit costs) and more thoroughgoing dispersal in time of crisis.

However, to argue thus is but to return to a wider theme. This is that truly radical solutions are urgently required to the danger of warplanes of all kinds being immobilised on the ground. No doubt the thinking that Western air forces currently engage in about this peril is positively revolutionary by any historical standards. Yet it is still nothing like enough of a departure to negate looming threats. Take, for instance, the Luftwaffe's submission for the Outline European Staff Target for a future fighter. This set 500 m as the maximum length of runway required either for take-off or for landing.[13] Similarly, the USAF requirement for a STOL demonstration variant of the F-15 envisages its relying on 1,500 ft of concrete for departure or return. However, it is all too easy to imagine that, come the exotic turn of this century, 10,000 ft runways could quickly be bombarded so comprehensively by pockmarkers, penetrators and scatterable mines that nowhere would there be that much footage intact. Moreover, were such a stretch to survive, it might be accessible overground only via sodden or icy grass that heavy machines might have great difficulty in negotiating. Furthermore, the hazards thus presented are among those that could be much aggravated by weather, darkness, EW, etc.

Granted, STOL capabilities of the sort just referred to might alternatively be utilised for operating off roads. Even so, a more fruitful line of development may be of machines that derive most of their lift from fixed wings but are able to take off or land vertically and, of course, to hover. This would be thanks to the installation of a horizontal rotor or tilt engines or vectored thrust or, indeed, 'X-wings' designed to be either fixed or rotating as occasion demands. The US Joint Services Advanced Vertical Lift Aircraft (JVX) development programme has been informed by seminal thinking along these lines. Never mind that barely a thousand such machines have been envisaged, over half of them for the US Marine Corps.[14]

All the same, it is not sensible to advocate VTOL without at least exploring dispersal. For one thing, it is an option that VTOL *ipso facto* presents. For another, forward dispersal eases the quite acute range constraints that VTOL labours under still. But the big difficulty remains that a developed distribution of one's air assets, especially when to improvised sites, does compound the command and logistic strains. Salient among the latter are those imposed by a continual demand for fuel. So the question that is raised is whether squadrons operating away from their established bases might not avail themselves of the civilian network for the resupply of automobiles. After all, even a village filling station may regularly stock something like 25 tons of petrol and diesel oil; and its counterpart along a highway perhaps a couple of hundred tons.

However, this question is not one that admits of an easy answer. Customarily, the assumption has been that the high performance demanded of warplanes depends on their being fed with fuel that is just right for them in terms of its specific density, calorific value and combustion efficiency. Furthermore, the switch within aviation as a whole from the use in piston engines of high octane petrol to the employment in gas turbines of kerosene (which is a rather heavier fraction of natural petroleum) is a change the military would be loth to reverse. This is because kerosene, thanks largely to its being less volatile, is less prone to uncontrolled ignition: a point confirmed, in a sense, by the preference shown for some of the most stable of the kerosene distillates in the constrictive yet exposed environment of the aircraft carrier. A further complication is that all planes likely to climb into or near the stratosphere need additives in their kerosene to ensure smooth vaporisation in the deeper cold. Besides all of which, engines always have to be tuned afresh if they are to accept a fuel substantially different from their normal one.

All in all, it is hardly surprising that the storage tanks on civil airfields, let alone anywhere else, have frequently been judged ineligible for military use. Nevertheless, there may be more scope for progress than at first appears. The several constraints upon civil–military interchange may be far less critical if (a) aircraft designers deliberately anticipate them; (b) some loss of engine efficiency is duly accepted; and (c) only low-altitude operation is envisaged. At all events, the approaching world shortage of petroleum is a stimulus to flexible thinking throughout this area, as is evidenced in the US Army's recent testing of methane as a helicopter fuel.[15] So, too, may be a prospectively vigorous development of the ram-jet in partial replacement of the rocket in tactical missiles. The connection here is that the attractiveness of the ram-jet solution may rest heavily on the creation of new fuels of higher calorific density.[16] Such work may then have wider applications.

Perhaps, too, certain ingredients could sometimes be fed into commercial fuel stocks on an *ad hoc* basis. At all events, some of the products of on-going research in this field (e.g., new additives for dampening down combustibility) will have clear civilian as well as military merit.[17] Some interest in the various possibilities is at last being sown among air staffs and industrialists, in spite of the inertia the 'air base' syndrome still tends to induce.

Arguably, the scatterable mine is the biggest single reason why the conventional warplane must give way considerably to the VTOL machine, be this monoplane or rotorcraft or hybrid. What can be said as well is that the multiple revolution in mine technology favours VTOL aircraft as against tanks and other vehicles given to overground movement. The main features of this revolution are the miniaturisation of sensors and also igniters; the further refinement of 'shaped charges', with their highly directional blast wave; the encasing of mines in plastic, a non-magnetic class of material; and fitting timing mechanisms to ensure self-destruction. Together they have made possible devices that may weigh a mere several pounds apiece yet still be devastating against wheels, tracks and vehicular undersurfaces. Admittedly, mines fitted with magnetic or acoustic sensors (plus perhaps some mechanism for rocket ejection) could alternatively pose some threat to aircraft flying very low. In the main, however, aircraft in flight are immune from this novel form of ordnance. Obversely, they can be appropriate instruments for its offensive dissemination.

Furthermore, there is a congruence between the argument in favour of VTOL and a no less compelling one in favour of a good proportion of an offensive air effort being devoted to Close Air

262 *A Multiple Revolution in War*

Support (CAS) more or less throughout a conflict. In part, the latter derives from military science notions, indicated in Chapter 5, about extended front operations in the context of dynamic non-nuclear war. Yet in part it stems from considerations which lie more within the realms of deterrence philosophy and political analysis.

To put them in a nutshell, it is nonsense to imagine that the West could abide aggressor armies penetrating at all deeply territories like Federal Germany or South Korea. The sheer mayhem and culture shock thus engendered would cause acute political, social and financial strains throughout those countries and further afield. Also, much loss of land would be crippling in itself, and would form an ugly backcloth for peace negotiations consequent on a cease-fire. While hostilities continued, the risk of the escalation of conflict to other theatres and into other forms (not least the renunciation of foreign debts) would be acute.[18] Accordingly, a pair of conclusions have to be drawn. The most fundamental is that the use of nuclear firepower, not to prosecute a war but to block its further development, is an option never to leave far in reserve. The most immediately relevant is that the cardinal aim must be to contain each enemy thrust at its very inception by a swift and sufficient application of non-nuclear firepower. Here CAS comes in with a vengeance.

Further to which, this rider should be added. The outlook for airborne troops as such may rest on their fulfilling two roles, each quite different from the mass descents on enemy territory so celebrated in the past. The first would be as broken down into relatively small raiding parties. These could be intended to disrupt either an attack or a defence in depth, as occasion demands. Then the second role would be to act as urgent reinforcements to unit positions or local sectors under attack. For both purposes, VTOL will often be the right method of force delivery partly because it subsumes, too, the possibility of force extraction. Here one may note that the airborne division which is to be found in the order of battle of the Bundeswehr land arm is one major formation that is largely dedicated to deploying for blocking actions across the line of Warsaw Pact armoured thrusts. Also worthy of remark is that the Spetsnaz raiding units now strongly developed by the USSR are believed to have NATO's nuclear installations as their prime objectives as and when they are dispatched as raider battalions or regiments.

Naturally, several other specific trends or prospects bear upon the dual emphasis here emerging: a VTOL capability primarily geared to the immediate support of ground operations along the FEBA. One is that men trained and equipped (especially with laser designators) to act as Forward Air Controllers may have to be more numerous within those units in contact with the enemy.

Next, there is that most vexed aspect of air–land battle management: the Identification, Friend or Foe of aerial vehicles. The present indications are that machines with revolving rotors will, in due course, lend themselves more readily than may straight monoplanes to positive recognition on the novel basis of their radar echoes;[19] and this should, on balance, make them still more eligible for full participation in dynamic war. Similarly favourable to the main line of argument is the aptitude VTOL machines possess (by dint of their ability to hover) for firing weapons from locations just inside their own FEBA and only a little above the ground. At present, this attribute is exploited as and when helicopters emerge briefly above trees, ridge lines, etc. to launch missiles against the flanks of armoured columns. But the likelihood is that, before the year 2000, certain VTOL models will be bearing their rocket-driven or ram-jet counterparts of self-homing artillery shells. These are the ones which, as they descend on a target area, will separate into several capsules, each able to guide itself onto a tank or whatever (Chapter 5). They will make fire from fully concealed positions more devastating than ever before.

NATO's Quest for Strategy

Undeniably, a divergence exists between the general thrust of military aviation as here envisaged by myself and recent doctrinal developments within the Atlantic Alliance as adumbrated in Chapter 2. Yet official concern with 'second-echelon targeting' and the like cannot here be written off as nothing more than bureaucratic obfuscation. All else apart, NATO as such has long taken a lively interest in future air warfare. Between its foundation in 1954 and a reconstitution in 1963, the SHAPE Technical Centre at the Hague was concerned only with this subject. Likewise in 1952, an Advisory Group for Aeronautical Research and Development (AGARD) was set up. This continues as an agency responsible to the Military Committee and publishes a rich variety of studies (see note 40, below).

Then again, no one could seriously argue that a strategy of Follow-On Forces Attack (FOFA) could prove appropriate in central Europe yet still be unsuitable for more general application. In other words, Europe is not a special case in that sense, quite the contrary. Only the most cursory examination of military geography is enough to show that, if such an operational strategy be viable apropos East Germany (where, after all, a very ramified road and rail network feeds an extended front line), then it must be even more so across the waist of Korea or around the Golan or the approaches to the Persian Gulf.

Therefore, the said doctrinal divergence has to be interpreted in other ways. The most elementary point to make is that, where the objects of air power are concerned, any sorting of priorities must be relative rather than absolute. No matter what doubts some of us may have about currently modish theories, nobody could advocate that the interdiction option be entirely foreclosed. After all, the disruptive effects of just a modest effort along these lines may be out of all proportion. Obversely, the other side has to divert a heavy weight of resources to anti-aircraft defence in depth, almost irrespective of whether that offensive effort be modest or huge. What is more, such weight may need to increase if tactical trends oblige air defence in general to become more distributive.

Besides, there are parallels between the interdiction of army columns etc. and attacks on military airfields, which is an operational requirement nobody is going to deny. Then again, no one is entitled, at least until the subject has been studied and discussed more fully, to have a final view about the dampening influence interdiction may have on the processes of policy determination and execution an adversary leadership will be engaged in. Those who believe this factor is never tangible would do well to ponder this. The day before the precipitate surrender by the Argentinian commander of the Falklands, a 1,000 lb laser-guided bomb had exactly and thoroughly wrecked one of his subordinate headquarters. Although this particular sortie must formally be classified as Close Air Support, the essential point can well be taken.[20]

Also, it is important to recognise that, in such matters, a staff planner and a professor do work to rather different time frames. True, each may be concerned in the round with the next quarter of a century or thereabouts. Even so, the former is obliged to think, first and foremost, about how best to use assets which exist today or may be available inside a decade. On the other hand, the latter has a duty as well as a disposition to stretch the purview, to reflect in the main on a more distant future.

Besides, the big rethink now under way in official circles emanated from various quarters around 1976 yet is still far from complete.[21] There are, in particular, the persisting doubts about the notion of any Warsaw Pact second echelon being sufficiently defined and sizeable to constitute a major strike objective. Reportedly, there is a school of analysts which says that Soviet staffs were once disposed to launch divisions into attack in successive waves but now tend to favour a single extended front.[22] Likewise, a debate continues about how the Operational Manoeuver Groups (see Chapter 2) are handled, the present inclination being to stress that they will most likely be committed to breakthrough and exploitation very early on.[23] Nor should we forget how much uncertainty also prevails about how much more thinly everybody's troops might be distributed around a future battlefield, and how well-defined the FEBA would prove to be.

Behind all of this may also be discerned some degree of cultural lag in respect of the shape and scale of the burgeoning revolution in non-nuclear firepower. A disconcerting precedent is the professional reaction to the machine-gun before 1914. Anybody at all cognisant of matters military knew well enough the rates of fire this weapon could register. Yet both doctrine and army structure continued to derive largely from the premise that assault *en masse* by infantry or even cavalry would always overwhelm it.[24] A corresponding danger today is that the future of air power is being worked out against a background of 1944 images of war.

What must also be reckoned with at present is strong political and public pressure in favour of 'raising the nuclear threshold' by exploiting new technologies. It is pressure which bespeaks to some of us a naive incomprehension of the nature of modern deterrence and warlike crises. This collective lapse has to be indicted not only for its intrinsic faults but because it feeds otiose distortions into the air power debate. Prominent among the latter is the further encouragement of an already overweaning doctrinal concern with strikes against rearward targets. Here the hidden danger is that this concern could lead to our side taking violent action ahead of any by the enemy. Anxiety on this score is perforce heightened by a repeated failure of advocacy in the professional press to exclude, in the explicit way it should, this pre-emptive option. No doubt one reason is that those involved are loth to compound existing disagreements about the disposition of aerial resources: disagreements already sharp enough, even as between — let us say — the USAF and the US Army.[25]

To which one must add this. NATO as such has long affirmed the precept that, whatever the circumstances, it will never initiate hostilities. Yet over the years, the West as a whole has inclined towards pre-emption (especially in the air) more often than is thought. Global pre-emption in the form of 'preventive war' (largely as waged by strategic air power) was on the US political agenda for a while in the early 1950s. Pre-emption by air, sea and land was resorted to collusively in the 'Suez' crisis of 1956; and then unilaterally by Israel in 1967 and again, in the Lebanon, in 1982. For 30 years past, NATO itself has had a declaratory strategy that allows for the first use of nuclear weapons. Now the choice in NATO Europe may lie between either continued acceptance of a lowish nuclear threshold or else a quasi-tacit dependence on pre-emptive violence in non-nuclear forms. Some of us prefer the former.[26]

Maritime Air Cover

At sea, a trend towards forms of organic air cover that are more distributive and based on VTOL should and probably will build up. Never mind the logistic and other disadvantages of operating aircraft off small ships. Never mind either the tendency for the limitations on range still associated with VTOL to be still more constrictive in the maritime domain. On both counts, technical innovations should prove helpful. For instance, it has been argued that a frigate of only 3,500 tons, built around a 'deep-vee' hull, could be made to operate five helicopters while on North Atlantic stations.[27] Meanwhile, exploratory work has been done by British Aerospace on whether a ship might use a special crane to retrieve and then hold a VTOL monoplane. Hopefully, this would enable vessels as small as 2,000 tons to replenish machines like the Harrier.[28]

Of no less consequence, however, is the likelihood that land-based air power will extend its authority across the High Seas more comprehensively than before. Trends in submarine pursuit and evasion are a major factor in this prospect. So is greater fuel efficiency in aircraft. So, above all, is the querulous status of modern surface ships as high-value targets too easily put out of action by air-delivered or air-directed weapons. In short, the vulnerability manifest so starkly off Norway in 1940 and Crete then Malaya in 1941 is aggravated by modern ordnance. It is doubtful whether the relative suitability of warships as platforms for close-engagement laser weapons could reverse this shift in advantage, even in the longer term.

Not that this prognosis is too discouraging to the West and its allies, blessed as they are with many long coastlines affording easy access to open ocean. Exploiting this advantage to the full may require, however, the development of a new genre of large aircraft with long range and endurance, these and other attributes enabling them to act as instruments of power projection to distant waters. Presumably, too, home bases in metropolitan territories or other locations well 'out-of-theatre' would not be regarded as licit targets, unless or until a conflict had become very bitter or extensive. Also, a virtual demise of the manned bomber in the strategic nuclear context will leave a legacy of tradition and experience eligible for application in this alternative way. Meanwhile, interest in a 'loiter' capacity is likely to be furthered by multiple advances in technology including those in closely analogous spheres. Among these may be a modest revival of the airship and the advent of surveillance drones (probably solar powered) able to remain aloft for days.[29] Evidently, too, the peculiar suitability of the maritime environment for the 'stand-off' delivery of weapons (especially at long range) also favours the cultivation of this theme.

A Mix of Systems

Yet however the future priorities of air power may be ordered in outline, years and years of very active experiment and evolution lie ahead as several S-curves move, at long last and more or less concurrently, into their acceleration phase. New modes of electronics will burgeon, not least as regards aerodynamic control. The purposive application to aeronautics of new materials will be of central importance. Loiter and VTOL will be chief among the capabilities aircraft designers will seek to realise, probably via diverse paths in each case. One such in the former may eventually be a revival of the flying boat. An oblique indication of the possibilities latent there can be seen in a recent Dornier proposal for a 1,000 tonne seaplane with a 400 tonne payload.

Interwoven with all of which will be continual dialectics about where the balance ought to be struck between, on the one hand, manned aircraft and, on the other, unmanned aircraft or surface-located systems or space-based ones. However, definitive judgements cannot be essayed in a study as extended in time span and as synoptic as this one aims to be. The electronic dimension is too uncertain and perhaps the human is as well. The geographical parameters will always influence the prospect. So will other contingent circumstances.

Still, it is safe enough to say that manned aircraft are unlikely to stay at all competitive against the various alternatives unless two principles suggested above are adhered to. The one is that cheapness must never be sought at the expense of effectiveness and survivability in the roles specified. The other is that airframes ought to be designed with more than one role in mind. As far as practicable, indeed, this dexterity should be preserved in every version of a given machine in service.

As the argument is taken further, the first comparison which invites itself is systems accuracy. Looking then at evidence from the contemporary West, one soon discovers not a lot to choose (at their respective maximum ranges) between artillery guns, strike aircraft, Ground-Launched Cruise Missiles (GLCMs) and surface-to-surface ballistic missiles. Even now, each can register Circular Error Probabilities (CEPs) which are to be measured in several tens of metres, the exact distance depending on the specifics of control and delivery. In theory, aircraft launching attacks close-in currently achieve the narrowest limits — meaning, on occasions, a CEP as low as 10 m. True, the differentials implied bear little upon wide-area ordnance: cluster bombs, napalm and Fuel Air Explosive. However, they could affect the application to hard targets of unitary warheads filled with high explosive. Then again, it would be wrong to overlook how big a bombload a plane may carry. Taken together, these two factors suggest a trade-off between outlay and benefits more favourable to the manned aircraft than is indicated by life-cycle costs alone.

But what allowance should be made for the human frailty of aircrews entering defended target areas? In all probability, a dichotomy occurs, one derived from the psychological mechanics described above. If a highly trained and motivated crew is launching one carefully planned attack against not too frightful odds, the mission they are executing may be far less prone to wild aberration than a similar foray with a Ground-Launched Cruise Missile (GLCM) might be. Thus the sort of assured precision achieved by the Israeli planes attacking the Iraqi nuclear complex in 1981 (every single bomb into the small target area) is not something one can yet depend on GLCMs for. On the other hand, repeated sorties by your average aircrew (often in darkness or poor weather) are sure to become rougher at the edges than purely technological prediction would lead one to expect. In other words, the idealised CEP of 10 m (only attainable, in any case, when laser guidance is to hand) will be subject to some upward revision. Besides which, flying machines with human operators on board will be less well placed to take the fullest advantage of the further improvements to be anticipated in terminal guidance, inertial navigation, etc. Yet the potential inherent in such developments, in terms of improved reliability and high standard accuracy, can be especially great for the pinpoint acquisition of targets.

So let us now turn to generic comparisons of the susceptibility of systems to interception. Cruise missiles hug the surface contours even more closely than do modern monoplanes. None the less, they are bound to be vulnerable to automatic fire directed at close range against their dorsal or ventral surfaces. Several years ago, certain US analysts were surmising that a heavy cruise-missile offensive launched against Soviet territory from Western Europe in 1985 would achieve only a 50 to 70 per cent penetration to the various targets prescribed, though a more advanced one in 1995 as much as 70 to 90 per cent.[30] In all probability, too, the initial 'learning curve' adjustment could be less irksome than in local defence against tactical aircraft, if only because both shock and EW would figure less.

What then of the other kind of missile, that projected by a rocket motor into an essentially ballistic trajectory? In US Army trials held as long ago as 1960, a Hawk SAM intercepted an Honest John travelling at Mach 2.5 on a trajectory some 20 miles across. Then in 1973, Egyptian aircraft released from above their own lines some twenty AS-5s: a Soviet air-to-surface subsonic rocket with a maximum range of something like 150 miles. Anti-radar versions did knock out two Israeli master radars.[31] Other damage was inflicted as well. As a rule, however, this slow and bulky device (length 8.6 m) fell easy prey to fighters and artillery, perhaps 80 per cent of those launched being shot down.

With the larger of the theatre ballistic missiles, on the other hand, the challenge presented to the defence is more like that posited by intercontinental rocket war. A Pershing 1A reaches Mach 8.0 in mid-course, which is nearly a half of what an intercontinental missile registers. What is more, the former's trajectory will be flatter in mid-course always, much flatter sometimes. Likewise, its flight time is never half as long. Therefore, stretching a SAM network to cope with Pershing or, of course, its Soviet counterparts could hardly ensure high defensive attrition, particularly in view of the part novelty would play in so high-speed an encounter. The most to hope for is that such a measure would complicate somewhat any adversary plans for a blanket surprise attack.

In reality, however, the most conclusive way to react to an onslaught of this sort might be to hit back forthwith at the launch sites from which the missiles have been observed to rise. Nowadays, any heavy weapons reliant on ballistic projection beyond the horizon (that is to say, artillery guns as well as rocket pads) lie exposed to the backwards extrapolation by enemy radar of the parabolic flight paths. Indeed, it is a state of affairs which affords a strong extra argument in favour of continuing to apply a good proportion of one's explosive firepower via aerodynamic platforms.

Among which, the RPV (plus perhaps its *alter ego* the drone) seems currently to enjoy an exceptional measure of immunity from defensive fire even when, as by the Israelis in the Lebanon campaign, it is dispatched with the intention of drawing this forth. Clearly, this is why some analysts in the West, NATO planners included, feel tempted to have these vehicles intrude deeply into adversary air space in search of tactical intelligence, even in situations of warlike confrontation that as yet fall short of actual hostilities. Yet as was noted in Chapter 2, such pre-emptive initiatives are bound to be read by the other side as menacing. What should now be added is that Soviet sensitivities on this score could well be heightened by two aspects of the technical scene. The one is that, for quite some time ahead, the USSR and her allies will be ill-placed to retaliate strictly in kind. Therefore they will face the choice between qualitative escalation and doing nothing at all. The other is that Russian military scientists have long been sensible of the operational value of electronic intelligence,[32] an input that unmanned intruders can make singular contributions to. Is not the moral that even this non-lethal form of pre-emption should not be embarked on until top-level deliberations have taken full account of the wider connotations?

A Multiple Revolution in War 271

Already, of course, RPVs and drones have proved technically suitable for a variety of reconnaissance missions. However, a comprehensive competence in wartime in this vital sphere has to involve as well data flows from satellites in orbit; manned monoplanes and rotorcraft; and the surface-located modes of surveillance. Apart from anything else, diversity of this sort is usually crucial to the assured collection of data in 'real time': an imperative more categoric than ever in the context of dynamic war, with many of the targets continually in motion the other side of the hill.

From all of which, this more general inference might now be drawn. Right across the spread of operational responsibilities that warplanes have assumed in our generation, they are destined to give way — steadily though not, at least in the foreseeable future, at all completely — to the other forms indicated. But in pursuing the question of how balances should successively be struck, the first matter to resolve always is relative costs over normal life cycles. Yet here caveats must be entered. Even allowing for modular replacement, particular systems will rarely stay adequately adapted for more than 15 years to a military environment that is ever more fluid. Then the next issue that ought to be addressed invariably (although it seems never to be at all) is how respective systems would compare, in the trade-off of cost and utility, if involved in actual war. So far as aircraft are concerned, it is often felt that there may be quite a pronounced 'knee' somewhere on the curve of rising cost plotted against rising effectiveness. Beyond it, the latter increases much more slowly in relation to the former.

In a sense then, the econometric approach is basic. Even so, almost nobody these days would accord it the ultimate importance it was rather skittishly conceded by some American academics for a while in the early 1960s. Too many unmeasured or contingent variables bear upon every evaluation, not least the prediction of losses. Nor can neat calculation encompass the likelihood that the future electronic ambience will alternate abruptly and continually between coherence and chaos. Nor can it ever hope to weigh the human factor, in all or even any of its aspects. Nor should our perusal be confined only to when a pitched battle is waging. Instead, the concept of a stand-back phase after a major encounter (as enunciated in Chapter 5) has to be considered.[33] Lastly, even if satisfactory formulae could be devised for each of these criteria, blending them together could only be by arbitrary weighting. Therefore, synoptic assessments can only be concluded by judgement and discussion.

Proceeding thus, allow me now to insist that the notion of a phase of 'stand-back' or suspended conflict does have import in this context. Not least does it as an argument for keeping more aircraft either in service or else in reserve than might have been justified on other grounds. As indicated earlier, the nub of the matter is that manned aircraft might come back into their own in a milieu already dislocated and denuded by a previous round or two of heavy fighting.

What is more, the above argument for according aircraft more priority than they might otherwise claim is underscored by this reality. Fleets of aircraft would take longer to rebuild than would, say, echelons of surface-to-surface missiles. This could in part be because new aircrew are peculiarly hard to recruit and train to tolerable proficiency. Unless the problem had been properly anticipated, however, the most acute constraint would probably be a shortage of the actual machines. Even if the industrial wherewithal is there to start with, an exotic modern warplane normally takes a couple of years to build. Nor should anybody imagine that a wartime sense of urgency could reduce this interval to anywhere near the several weeks achieved with, say, Battle of Britain fighters. Besides which, a spell of heavy fighting would probably have increased the now standard excess of crews over planes.

Air Power and Geography

Against this background of complexity and flux, the best definition of 'air power' for tomorrow may be one which is studiedly naive and tautological. Namely, it is that *dimension in twentieth-century war that has customarily been provided very largely by the manned warplane*, no matter that this instrument *per se* will have to give way considerably to alternative systems. Working then to this formulation, it is not immediately obvious that the land-based 'air power' of the West and its allies need contract all that much in the foreseeable future. Arguably, it will not have to contract at all. Perhaps all we should seek is a progressive revision of aim and configuration, partly in order to provide more in the way of (a) tactical mobility in support of the land battle; and (b) air cover out to sea.

Still, these are not matters to resolve for good in 1986. The future of air power must henceforward be kept under continual review. It

A Multiple Revolution in War 273

must be in pursuance of stable military balances but also because open debate is the best way to ensure that the imperatives of technological change do minimal damage to service ethos and morale. Above all, we need regularly to ask what the spread of effort is and ought to be between the three recognised dimensions of war. Unfortunately, however, about the only evidence we have derives from the budgets and manning levels of what are still, in like fashion, very generally accepted as the three armed services: army, navy and air force. Yet, bearing in mind what has been said above, this is less than satisfactory as a basis for delineation.

For what they may be worth, however, some representative figures are as follows. In 1964, according to the International Institute for Strategic Studies, the air forces of the European members of NATO had an active manning strength of 550,000, as against 2,075,000 in the various armies: a ratio here of 1 to 3.77. In 1979, the balance was given as 497,000 versus 1,774,000: 1 to 3.57. In 1984, with Spain in NATO, it was 546,000 versus 2,146,000: 1 to 3.93.[34] In short, no trend against air power currently obtains. As much is confirmed by defence expenditures per serviceman, taking as typical enough the British experience cited in Chapter 12.

Two comments invite themselves. The first has again to be that the new spirit of toleration between the three services (evident in most countries in recent years) may well encourage undue rigidity in these matters. But the second is that, were the fraction of resources so constantly devoted to allied air forces diverted instead to armies, a thickening up could occur on the ground that in some places (e.g., Germany and Korea) would be most valuable. Therefore, it is imperative that air power justifies itself, first and foremost, by its direct contribution to any surface battle. Across the years, alliance air staffs have broadly accepted this. They have done so by giving highest priority to Close Air Support, even in the RAF. This pride of place has in NATO been manifest via internal doctrine; the structuring and management of airborne resources; and a special network for tactical control.[35] So the burthen of my argument is that this ordering of things should not now be overturned. Rather it should be accentuated, albeit in new forms.

Irrespective of where emphases are placed, however, the surface-to-space continuum observed in Chapters 1 and 5 is starting to become a seamless reality *vis-à-vis* not only surveillance but also target destruction. In the one direction, there is the Space Shuttle and the USAF's experimental use of an F-15 fighter to launch a missile against a satellite in orbit. In the other, there is the surface-to-surface missile, winged or otherwise. There is also what needs must be a marked expansion of air mobility on the battlefield. As regards which, strong managerial justifications can be adduced for any extra capacity being made the responsibility of the air arms proper, a point well taken in one AGARD report on the flying of helicopters.[36] Nevertheless, armies will always expect all those who fly the VTOL machines to share more fully than aviators have traditionally done the soldierly perspectives on battle. So much for psychological detachment and 'Antiseptic War'!

Does this mean, then, that air arms everywhere should cease to be independent services? The reply ought surely to be negative. All else apart, you do not waft away atavistic conservatism by fitting everybody out in uniforms dyed the same colour. Instead, you may recharge it. In the past, a monochrome connection has not precluded antagonisms between, say, cavalry and infantry or even horsemen and tankers. Nor has it moderated the feuding between battleship and carrier or bomber and air defence lobbies.

Besides, the attitudes nurtured by air forces may stimulate constructive change. My own impressions as an Academic Consultant, from 1972 to 1978, to Britain's National Defence College, then at Latimer in Buckinghamshire, were essentially as follows. Despite a rich accumulation of operational and political experience throughout our long years of imperial withdrawal, the soldiers tended to be too diffident about matters of high policy, force structure included. Meanwhile, the Royal Navy too wilfully remained the distant 'Silent Service'. The RAF, on the other hand, was gripped by an awareness of being the youngest of the three. It was acutely conscious of having to hold its own by dint of argument in the corridors of power and influence. Occasionally, the manifestations thereof were too partisan to carry conviction. As a rule, they were not.

Some years ago, the nub of the case for keeping a triadic organisation was put with sardonic felicity by a then Deputy Assistant Secretary of Defense in the Pentagon. He contended that, since manned aircraft are no longer unique as a:

> means of getting an explosive charge from here to there . . . the only reason for having separate air forces has . . . disappeared. But we have them and therefore should plan to use them as effectively as possible. They do seem to be somewhat less rigid than

the other services in their thinking. In my country, for example, the Air Force is in the forefront of developing new ways of doing things.[37]

What he was referring to was a more positive attitude to technocratic modernisation and, in some measure, enlightened personnel management. In the West at any rate, these traits do not correlate (except sometimes in a negative sense) with more 'progressive' political attitudes, however that term be interpreted. Academic analyses — from Sweden[38] and the United States — have confirmed that air forces do not present 'a picture of decisive deviation from the basic conservative orientation'[39] of the military. On the other hand, a fervent commitment to mobile defence, in line with the precepts of the ill-fated Marshal Tukhachevsky, may account for the corps of air-force officers suffering exceptionally severely during Stalin's savage purges of the Soviet military in the late 1930s.[40]

All in all, we probably ought to characterise future air power in a manner akin to F. W. Lanchester's. In fact, the kind of nuance that might best be struck will perhaps be indicated by some further consideration of 'ubiquity', the characteristic so often presented as a singular and supreme attribute of air power. Though extolled by the early apostles of the air, it was to be celebrated even more as the advent of ultra-long-range bombers brought about the geostrategic revolution of intercontinental exchanges across the north polar wastes.[41] For a while in the 1950s, this scenario was closely linked with Dullesian notions of 'massive retaliation' at 'a time and place of our choosing' against Soviet or Soviet-inspired probes anywhere around the Eurasian rimland.

Nowadays these notions are aired but spasmodically. None the less, the view of strategic geography that underlies them gains credibility from the technical possibilities ahead of us. As was remarked some time ago by Richard Burt (now US Ambassador in Bonn), 'a terminally-guided cruise missile with a conventionally-armed earth-penetrator warhead could be used versus hardened silos or command centres in the Soviet Union'.[42] Such finesse across such distances could, of course, offer useful early rungs in a ladder of controlled escalation designed to force a stand-off in some local conflict. Likewise, much scope exists for the aerial projection of power across extended sea areas. In some respects, too, a VTOL aircraft is as ubiquitous a device as can be imagined.

Considered in the round, however, things may not be as straightforward as that. Take that most familiar and crucial of parameters, the destruction of manned aircraft as they are engaging in offensive sorties. This is still liable to vary widely from scene to scene. It is so largely under the influence of such circumstantial factors as the quality and suitability of the defensive systems, the broader electronic milieu, and the square law. In Indo-China, the USA faced a relatively benign environment in these particular respects. Thus the defenders of North Vietnam had to fire a hundred SAM-2 high-altitude missiles for every American aircraft thereby brought down; and although it seems that every single US helicopter in the field in the South was fired on at least once during the climactic Tet offensive of early 1968, the loss rate thereof was only one per 1,000 sorties.[43] At another place or time, however, the West might face much tougher opposition, necessitating very different judgements about the weight of air power required; the part it could play; the relationship between warplanes and alternative means to the stipulated ends; and the optimum designs for individual systems.

By the same token, such background factors as topography, vegetative cover, the availability of airfields and ports, human settlement patterns, the ratio of troops to space, and the run of weather will continue to affect the fortunes of air war. Indeed, their influence may be accentuated by the delicacy with which the balance of advantage may be struck in future exercises of air power. In other words, the specifics of geography could shape the air–surface interaction even more in, say, 1995 than half a century before. But geography does not repeat itself.

References

1. Neville Brown, 'Air Power in Central Europe', *World Today*, vol. 39, no. 10, October (1983), pp. 378–84.
2. See *Military Balance, 1984–1985* (International Institute for Strategic Studies, London, 1984), pp. 17–21.
3. See the author's 'The Myth of an Asian Diversion', *RUSI Journal*, vol. 118, no. 3, September (1973), pp. 48–51.
4. M. J. Armitage and R. A. Mason, *Air Power in the Nuclear Age, 1945–82* (Macmillan, London, 1983), pp. 2 and 226.
5. F. W. Lanchester, *Aircraft in Warfare, the Dawn of the Fourth Arm* (Constable and Co., London, 1916), p. 145.
6. William H. Rees, 'The McDonnell Douglas Dual-Role F-15 Programme', *Military Technology*, 4 (1984), pp. 14–21.
7. *Flight International*, vol. 126, no. 3924, 8 September 1984, p. 538.

8. This Soviet expression means much the same as the NATO term, Forward Edge of the Battle Area (FEBA). See Chapters 2 and 5.
9. FM 100-5 *Operations* (Department of the Army, Washington DC, 1976), ch. 8.
10. Agardograph 233, *Assessing Pilot Workload* (AGARD, Neuilly-sur-Seine, 1978).
11. O. G. Sutton, *Understanding Weather* (Penguin, Harmondsworth, 1969), chs. 5 and 6.
12. John Dalling, 'Testing the Mobile Sector Operations Centre', *Air Clues*, vol. 38, no. 12, December (1984), pp. 450–1.
13. *Military Technology*, vol. VIII/5 (1984), p. 41.
14. L. Kim Smith, 'JVX: a tilt-rotor aircraft for the US Services', *Military Technology*, vol. VII, no. 1 (1983).
15. *Aviation Week and Space Technology*, vol. 120, no. 3, 16 January 1984, p. 135.
16. Brian Winstall, 'Integral rocket ramjets for long legs at high speeds', *Interavia*, 12, December (1984), pp. 1331–4.
17. *Daily Telegraph*, 8 November 1984.
18. Neville Brown and Anthony Farrar-Hockley, *Nuclear First Use* (Buchan and Enright for RUSI, London, 1985), Part II, ch. 1.
19. 'Radar Classifies Helicopters', *Flight International*, vol. 125, no. 3900, 4 February 1984, p. 343.
20. Jeffrey Ethell and Alfred Price, *Air War, South Atlantic* (Sidgwick and Jackson, London, 1983), p. 207.
21. See, e.g., Major-General John W. Woodmansee, 'Blitzkrieg and the Airland Battle', *Military Review*, vol. LXIV, no. 8, August (1984), pp. 21–39.
22. 'Fee, fi, fofa, hm', *The Economist*, 24 November 1984.
23. For a pair of well-informed articles on this subject see *Armed Forces Journal International*, vol. 122, no. 1, August (1984), pp. 51–60.
24. Brian Bond in Michael Howard (eds), *Theory and Practice of War* (Cassell, London, 1965), pp. 97–125.
25. Major James A. Machos, 'Tacair Support for Airland Battle', *Air University Review*, vol. 35, no. 4, May–June (1984), pp. 16–24.
26. Brown and Farrar-Hockley, *Nuclear First Use*, Part II, chs. 3 and 6.
27. E. H. Serter, 'An Air-Capable Frigate for NATO', *International Defense Review*, vol. 17, no. 8 (1984), pp. 1114–15.
28. David A Brown, 'Sky Hook to help Ships launch Harrier', *Aviation Week and Space Technology*, vol. 113, no. 3, 6 December 1980, pp. 105–7.
29. 'Long-Endurance Drones Mature', *Flight International*, vol. 126, no. 3937, pp. 1567–9.
30. *The Economist*, 20 February 1982.
31. Armitage and Mason, *Air Power in the Nuclear Age*, p. 134.
32. See e.g., *Principles of Jamming and Electronic Reconnaissance* (Military Publishing House, Moscow, 1968), Introduction and ch. 10.
33. Brown and Farrar-Hockley, *Nuclear First Use*, pp. 63–4.
34. These figures are culled from successive editions of *The Military Balance*.
35. Lt. Col. D. J. Alberts, USAF, *The Role of Conventional Air Power* Adelphi Paper 193 (The International Institute for Strategic Studies, London, 1984), p. 21.
36. Advisory Report No. 69, *Helicopter Crew Fatigue* (AGARD, Paris, 1974), p. 9.
37. John H. Morse, 'Advanced Technology in Modern War', *RUSI Journal*, vol. 121, no. 2, June (1976), pp. 8–12.
38. B. Abrahamsson, 'The Ideology of an Elite: Conservatism and National Security', in J. Van Doorn (ed.), *Armed Forces and Society* (Mouton, Hague, 1968), pp. 71–83.

39. Morris Janowitz, *The Professional Soldier* (Macmillan, New York, 1964), p. 238.
40. M. J. Deane, *The Political Control of the Soviet Armed Forces* (Crane Russak, New York, 1977), p. 43.
41. For a useful adumbration see Saul B. Cohen, *Geography and Politics in a Divided World* (Methuen, London, 1964), pp. 49–51.
42. *Survival*, vol. 18, no. 1, January/February (1978), p. 15.
43. John Clements, 'Air Defence Mythology', *RUSI Journal*, vol. 127, no. 3, September (1982), pp. 27–32.

APPENDIX A: THE ELECTROMAGNETIC ENVIRONMENT

The keys to a basic understanding of this ever-more important milieu of peace and war are provided by Quantum Theory: the corpus of thought evolved during the first 30 years of this century, principally to help explain the behaviour of energy and matter at the smallest limits of what can be observed or inferred. One of its findings has been that, at such minuscule dimensions, particles behave like waves and waves like particles. Thus the successive waves in an electromagnetic emission can usefully be seen as assemblages of discrete parcels of energy known as photons. What is more, a precise and inverse relationship is perceived between photon size and length of wave: the greater the former, the shorter the latter.

Another cardinal precept is that a similarly inverse and immutable relationship exists between the frequency and the length of any electromagnetic waves. As indicated in Table 4.1, the wavelength in metres multiplied by the frequency in megahertz (or millions of cycles per second) works out, to all intents and purposes, at 300. This is because the universal constant always referred to as the 'velocity of light' is slightly under 300 million metres a second.

Duly, various wavebands are identified in terms of their spread within the electromagnetic spectrum, again as in Table 4.1. Often, that part with wavelengths not below one millimetre is spoken of as the 'radio range'; and that portion of the radio range with wavelengths not longer than one metre is normally encompassed by the term 'the microwaves'. A looser though common practice, at any rate with 'passive' sensing (see below), is to describe all wavelengths below 1 m as 'infra-red'.

Frequency and wavelength are interchangeably used to denote any point on the electromagnetic spectrum from the VLF through to and including the EHF. In the remaining segment, however, reference is normally made to wavelength alone. A related convention is that whereby, from Low Frequency to EHF inclusive, each broad sector identified embraces an order of magnitude (factor of ten) alteration in frequency/wavelength. But at the shorter wavelengths this practice, too, is set aside, one reason being that 'visible light' is a function of human adaptation not mathematical symmetry.

Still, this standard format (exemplified in Table 4.1) has been subject to variation. Since World War II, government agencies, manufacturers and so on have successively devised lettered codes, the intention being to indicate better those parts of the spectrum that radar regularly employs. Among the more radical departures has, in fact, been an EW formulation lately adopted within the US military and other parts of the alliance. The high frequency limit of the VHF is withdrawn from 300 megahertz to 100; and above it on the frequency scale comes a new series of bands of irregular width, these being delimited strictly by frequency and designated alphabetically from A to M. Most modern radars fall within the ambit D to J inclusive — that is to say, between 1 and 20 gigahertz.[1]

One truism that readily emerges from any such tabulation is the tremendous spread of wavelengths within the ambit of operational concern. To be more explicit, the shortest VLF wavelength is no less than ten thousand million million times as great as the longest cosmic ray one.

Before examining the implications of that spread, however, it is necessary to delineate the prime divide in electromagnetic sensing techniques. It is that between passive and active: the former being the kind that does not emit radiation itself, interpreting instead emissions from the target area. It saves both money and weight and does not betray itself by dint of being switched on. Chief among its drawbacks is an inability directly to ascertain range, something an active sensor does by measuring the time lapse between the transmission of a pulse and its return.

Still, two passive sensors may sometimes be able to locate a target approximately by triangulation of the respective bearings on which they are receiving emissions. Also, work is being done on the passive inference of range from the time lag between the arrival straight from an emitter of a recognisable pulse and its arrival via reflection off the ground. But an obvious difficulty about this technique is its dependence on correct assumptions about the altitude of the emitter and the geometric symmetry of the reflection.

Furthermore, an element of good fortune in recognising the characteristic emission of the sensor's target is a prerequisite in all passive modes. By contrast, active radar has the advantage that it determines the frequency time and bearing of a pulse that will in any case, have to be made 'coherent'. What 'coherence' here means is that all the waves are in phase, with all their lengths effectively

Appendix A   281

identical or, at any rate, in close and regular relationship.

In fact, the application of this concept is utterly fundamental to the conduct of Electronic Warfare (EW). For one thing, the higher the frequency, the harder it is to generate radiation coherently. Think of the great advantage Allied air and naval power is always deemed to have derived from the invention late in 1940, at the University of Birmingham, of the cavity magnetron: a means of effecting the generation, in small spaces with low power inputs, of coherent rays in the UHF range. Then in 1960 came a further momentous invention, the LASER: Light Amplification by the Stimulated Emission of Radiation.

Lasering can be defined as the generation of coherent radiation in or near the visible-light sector. To initiate it, energy is 'pumped' in from outside. A proportion of this will cause certain electrons revolving round atomic nuclei to 'jump' from a lower to a higher level of energy: a general shift known as a 'population inversion'.

If then an electron thus brought to a higher energy level is stimulated by a photon of sufficient size, it will jump back to the lower, shedding a photon as it does so. Similarly, this photon may stimulate another such electron, thereby liberating a further photon of the same size. A 'chain reaction' or 'avalanche' of identical photons will thus be initiated. What is more, the resultant stream thereof will be coherent. The conceptual explanation of this phenomenon is to be found in aspects of Quantum Theory. The most serious practical snag, as things stand today, is that the resultant energy output will be 5 to 10 per cent of the power input. Across much of the electromagnetic spectrum, in fact, one very consistently finds that the shorter the wavelengths being sought, the harder it is to generate radiation distributed with adequate power just around the lengths in question.

Alongside this discouraging aspect has to be set another characteristic that can likewise be explained by a strand within Quantum Theory, this particular one being the Uncertainty Principle enunciated by Werner Heisenberg in 1927. Among its more important tenets is that photons, electrons and the like are governed not by certainties but by probabilities. One consequence is that, even were one to attain the unattainable ideal of perfect instrumentation, no beam of electromagnetic energy could ever be made perfectly parallel. Divergence will always occur as successive wave fronts advance. In other words, there will always be a steady decrease in energy with distance.

However, this angular spread is in a direct linear relationship with the wavelength employed. To halve the latter, say, is to halve the former. Obversely, there is an inverse relationship with the width of the beam at source. Thus if the initial width is doubled, the subsequent spread is halved. So the essential reason why an anti-aircraft searchlight, say, is able to project a beam that appears to the casual observer not to diverge at all is that its paraboloid reflector is very wide in relation to the mean wavelength of visible light. With antennae in other configurations (dipoles, oval slots in a metal plate, and so on), the ratios between the structural dimensions and the wavelength remain important.

So relating these variables via a simple formula,[2] one finds that a parallel beam projected at, say, 30,000Å via an aiming mirror 10 cm wide will diverge by one part in 10,000 or 10 microradians. In other words, over a distance of 10 km its width will increase by a further 10 cm, a broadening that is appreciable but may not be incapacitating.

Another merit of the shorter wavelengths, as applied to surveillance, is their potential ability to discern smaller images and give more exact definition to the rest. For should any two points from which wave energy is being emitted or reflected be closer together on a given bearing than are similar positions on two successive cycles of the wave front, those points will not be distinguishable one from another, irrespective of the quality of the sensoring system.

Correspondingly, a prime consideration with telephony or telegraphy is that the concentration of channels that may, in principle, be possible rises with the frequency. Thus the saturation capacity for high-speed telegraphy in the VHF is, even in theory, only 50,000 channels. Yet, save for all signals being prone to wander somewhat from the frequency selected for them, the SHF could accommodate 50,000,000.

Classically, a basic test of electronic superiority has been an ability to operate (actively or passively; and with adequate power) on the higher frequency. This has partly been in view of the attractions of utilising reaches of the spectrum an adversary cannot so readily attain. But it is also due to the inherent attributes just alluded to.

Accordingly, this frontier continues to be pushed forward, as is evidenced in the burgeoning application of millimetric radars (in self-homing devices, terrain avoidance, etc.) and, of course, of lasers. Taking the situation overall, however, a comparison of high frequency operation — even assuming it can be made in a meaningful fashion — may count for less these days than does the constant diversification and refinement of other EW stratagems. All else apart, modern analysts are very conscious of the complications that operation on the higher frequencies may pose. Thus one technical problem that presents itself in the construction of microwave transceivers is that energy has to flow to and from the antennae not through cables but through hollow tubes called 'waveguides'. The maximum wavelength such a tube can properly transmit is determined by its cross-sectional dimensions. Yet by much the same token, any operation on a wavelength shorter than the one or two to which its cross-section may be tailored will occasion an appreciable loss of efficiency, except where the former can be an exact divisor of the latter. Granted, a measure of extra agility may be secured by fitting alternative sets of waveguides. But that means greater complexity.

Then again, complications can be caused, in various parts of the microwave sector, by an unduly high prevalence of harmonics, which are emissions in phase with — though a number of times shorter than — the primary wave. More seriously, there are several distinct respects in which the higher frequencies may be disadvantaged near the Earth's surface. Thus neither the microwaves nor, in fact, any shorter wavelengths lend themselves at all well to transmissions beyond the horizon. For unlike the longer waves, they are not subject to reflection from the ionosphere: that zone between forty and several hundred miles up within which the very thin atmosphere present has been considerably ionised — in other words, many of its atoms are separated out into nuclei (positively charged, as always) and free electrons, each with one negative charge. True, there is the positive corollary that microwave radar can peer through the ionosphere to observe warheads, satellites and — of course — many other entities, near and far. Correspondingly, a myriad of radiations from distant galaxies, space vehicles in orbit, and so on penetrate to the Earth's surface. In the military context, however, one has to enter the caveat that nuclear explosions within the zone in question can block very extensively (and for minutes or even hours) waves at normal radar frequencies and, for that matter, across much of the electromagnetic spectrum. Even within the lower atmosphere, indeed, the Electro-Magnetic Pulse (EMP) that is part and parcel of a nuclear explosion can interfere seriously with electronic systems in the general vicinity.

A further dimension of this theme is that certain experimental radars which operate in the HF band may exploit 'ionospheric bounce' to detect Over-The-Horizon (OTH). The US Advanced Projects Research Agency has worked on OTH radar since 1962; and for several years past, there have been indications of a corresponding Soviet programme.

However, nobody denies that OTH has serious weaknesses. Given that the shortest wavelength in the HF band is 10 m long, the discrimination achieved will be much too coarse properly to discern, say, individual fighters in close formation even though it should be able to pick out, as a single 'resolution box', the formation itself. It is too oscillatory and uneven at the best of times, and also too subject to aurora and related disturbances. Besides which, geodetic considerations render ineffective at ranges of less than 1,000 km OTH of the 'backscatter' variety, the sort in which the transmitter and receiver are co-located, more or less as in ordinary radar. Never mind that this mode does otherwise seem more efficacious than 'forward scatter', the version in which transmitter and receiver are positioned far apart. To all of which must be added the operational vulnerability of OTH, the antennae arrays inevitably being several hundreds or several thousands of metres across.

All things considered then, this form of radar ought to be discounted in tactical air warfare overland. Quite possibly, it could help to detect the long wakes of ionisation caused by ascending ICBMs. Perhaps, too, it will eventually contribute materially to the tracking of strategic air movements or distant naval task forces. Yet even on these scores, scepticism is induced by the halting progress of OTH-B: the USAF programme whereby two backscatter complexes (one in Maine and the other in the Pacific North-West) will flank the Distant Early Warning screen across Canada, thus extending the compass of surveillance of the approaches to the continental United States. Likewise, a sustained Australian interest in OTH has to be related to the oceanic and continental parameters of Australia's defence and also to its surveillance being mainly oriented well away from the south magnetic pole.

Still, the statement that the microwaves have only a line-of-sight application is not entirely adequate. Abnormal temperature or humidity changes with height can cause ducts that, however locally and temporarily, may increase ranges several times over. Furthermore, it has for 40 years been realised that even the lower troposphere can,[3] via a mechanism not yet fully understood, reflect or refract small percentages of microwave radiation, thereby allowing low-intensity communication across a few hundred kilometres. By 1965, indeed, NATO had invested $100,000,000 in a chain of 'tropospheric scatter' stations from the north of Norway to eastern Turkey; and discussion, couched in quite an optimistic vein, of the potentialities of this technique has lately taken place in the Soviet military press. Meanwhile, tropospheric scatter has established a secure place for itself in the field of civil communications. Over and beyond all of which, the progressive introduction of efficient relay stations in the form of artificial Earth satellites means that the phrase 'line-of-sight' need not always have the constrictive connotations it formerly did.

Reverting to beam divergence, it is needful to enter a caveat. It is that narrowness is not always an unqualified virtue. A ground-to-air search radar, for example, may require a beam that, while well-concentrated in the horizontal plane, is wide in azimuth. Then again, if the beam actually tracking a bomber or guiding a missile has too small an angle of divergence, it may too readily lose its target. On the other hand, the narrower beams may be especially well adapted to picking out objects which are stationary or slow moving. A simple pulse radar with a beam as wide as, say, two degrees that received an echo from, say, a ship hove to 50 km away might well be unable to identify this echo against the background of clutter from something like 25,000,000 sq. m of water. This is because the total image projected by this expanse might well correspond to that from a perfect reflector perhaps 1,500 sq. m in notional extent and presented perpendicularly to the said beam.

For workaday purposes, the edge of a beam is actually defined as being the surface along which the intensity of the emission is half what it is at the maximum. In reality, however, this gradation is normally a lot more complex that this definition suggests. For any beam projected at all near to a ground or sea surface is liable to comprise a number of 'lobes' of high power separated by 'nulls' of low power, the thrust and configuration of the former depending on antenna design.

Which design is chosen will be governed largely by whether the prime aim is to project a narrow beam or a powerful one. For while a narrow main lobe may well serve as a basis for a narrow main beam, it may tend also to involve a comparatively large fraction of the total emission being channelled through side lobes which will represent continuous losses of power and act as sources of spurious

286 *Appendix A*

echoes. Conversely, if as much of the total effort as possible is canalised into the main lobe, that lobe may then prove strongly divergent. When one is talking in terms of lobes rather than beams, the half-power level is still what is used for boundary definition.

Yet no factor in these off-plays between different modes and different frequencies is more consequential than the way the passage of electromagnetic waves through any substance is subject to attenuation by diffraction and absorption. Seawater, in particular, is so consistently opaque that long-distance communication with fully submerged submarines is only possible by means of extremely large and powerful surface-based transmitters in the VLF range; and not even with them unless, in fact, the vessels in question are trailing aerials to within 20 metres of the surface. Absorption apart, VLF is useless for detection purposes because its shortest wavelength is 10,000 metres.

Then there is the extensive interference, especially with radar transmission, that may be caused by aggregations of water droplets or ice crystals suspended in the atmosphere as fog or cloud. Also, of course, by actual precipitation. In addition, however, the atmospheric gases (water vapour, oxygen, carbon dioxide etc.) absorb electromagnetic waves considerably. The overall patterns are complex but there is an underlying tendency for absorption to get ever more marked as frequencies climb through the SHF and beyond. Thus a peak caused by water vapour is encountered at the SHF wavelength of 1.35 cm; and this accelerates the decline of intensity with distance by five per cent a kilometre. Likewise, low-power EHF wireless is used to link warplanes with their aerial tankers, this for the very reason that such intercommunication is unaudible to any third party more than a very few thousand feet away. At wavelengths below 1,000Å, in fact, there are few good 'windows' between the successive absorption bands.[4]

The peculiarly negative impact on passive sensing becomes evident through an examination of the inverse relationship between a given body's temperature (as measured on the absolute or Kelvin scale[5]) and the wavelength of peak intensity of the electromagnetic emissions that are occasioned by its warmth. The peak 'black body' emission from a land or sea surface at the not untypical temperature of 300° K will be *c*. 1,000Å. The leading edge of the wing of a monoplane in flight might well achieve a temperature of 500° K; and if so, the peak emission will be 580Å. From the flame of a gas turbine at 2,000° K, it will be 145Å. The temperature of the Sun at the surface is around 6,000° K. Therefore, 50Å will be its emission peak. What this indicates, of course, is that infra-red discrimination of, say, the hot exhaust pipe of a jet engine is much easier at very high altitudes than lower down. For as the atmosphere thickens, absorption in the most relevant frequency range sharply increases. This means that (a) the effective ranges for detection will be reduced, and (b) it will be harder for the sensor to distinguish between a jet exhaust or some less vital aircraft segment or a decoy flarepot or even the Sun.

Yet to note this is not to discount the manifold advances now being made with 'infra-red', as broadly defined. Nor is this progress all that surprising. Only 20 years ago, it was plausibly being asserted that cumulative expenditure on 'infra-red' by Western countries had been hundreds of times less than on radar. In the interval, this imbalance has been appreciably reduced.

Not that radar has reached as yet the asymtotic limits of its conceptual development, not to mention multifarious application. In reviewing the outlook in this area, it may be appropriate to start with what may be termed the archetype — the radar that sends out identical pulses of energy at fixed intervals. Since the spread of wavelengths within any of these pulses is so small, it is customary to speak as if just the one point on the spectrum were involved. Nevertheless, the exact extent of this spread is a matter of some practical consequence. Should it be a bit too narrow, the received signal may lack definition. Should it be too wide, the emission becomes easy to locate and jam. Similarly, the duration of a pulse is relevant. Usually, this is measured in microseconds — millionths of a second — and can be anything from a quarter to several hundred thereof. So here again the choice made will be a compromise between competing aims. While too little energy to override background or adversary interference may be released in a short pulse, a long one will be inimical to good target definition and accurate range registration. Furthermore, the archetypal form of radar cannot determine range when the transmission of one pulse precedes the return of the previous one. These days, however, a set may be so designed that it can subtly vary the median frequency while a pulse series is being transmitted. Concurrently, it will separate out the frequency groupings of the returning echoes in order to time them independently of each other.

Another constraint to reckon with is that the power demanded in a system specification may tend to outstrip the feasible input. So not infrequently, the 'duty factor' — the ratio by which the interval between the commencement of two successive pulses exceeds the duration of each pulse — has to be set as high as 10,000.

Nevertheless, experiments were under way by 1945 with radars that produced Continuous Wave (CW) rather than pulses. Such a set is inherently well adapted to having its frequencies so modulated that range can be calculated from the time taken for a particular frequency to return. Alternatively, the CW frequency may be kept steady in order to take advantage of the doppler effect. By this is meant the shift that takes place in the perceived frequency of a wave propagation when it is reflected off a moving object, a shift that can serve to differentiate a low-flying aircraft from the land or sea beneath and also afford a measure of its velocity relative to the observer's. These days, CW is often broken down into a beat sequence, so that both the pulse and the doppler principles can be exploited.

One strictly geometric axiom is that, were reflection perfect and if no absorption occurred *en route*, the magnitude of the received echo would be inversely proportional to the fourth power of the range. In other words, a doubling of the range will reduce the echo by 15/16ths and a trebling by 80 parts in 81. This is because an advancing wave front will represent all or part of an expanding sphere; and so the surface it is spread over grows in proportion with the square of the distance from the point of origin. Likewise as soon as this front echoes off the target to commence its homeward journey, its intensity becomes subject to the inverse square law all over again. The fourth-power law is often spoken of, rather colloquially, as the 'radar equation'.

Add to it the physical reality that, within the atmosphere, significant absorption always occurs *en route*; and also that no target surfaces are perfect reflectors, most being very prone to absorb, deflect and scatter. Then the upshot is that the radar pulse received is liable to be a most diminutive version of that transmitted. Thus an air search radar with a peak output of perhaps one megawatt (a million watts) might have to handle a peak return of but a few microwatts, or millionths of a watt.

This state of affairs aids marvellously those who seek to deploy, against a given radar, Electronic Warfare in the form of jamming or deception. Yet what may, even in a benign environment, appear the impossible task of registering and interpreting responses so feeble has been made a lot easier, not least apropos large surface-based systems, by several technical developments. The most universally applicable has been the virtually complete substitution (starting from 1954) of the semi-conductor for the vacuum valve, the benefits therefrom including compactness, ruggedness, less power demand and a big reduction in the internal generation of the cacophony of unwanted signals usually spoken of as 'noise'. Further diminution of such interference may be achieved, particularly in the larger and less mobile systems, by cryogenic cooling of the critical components. Granted this improvement may not be worth the trouble where the ambient or external noise level is bound, in any case, to be high. But the balance of the argument may be different where the external contribution is likely to be low. Scanning the sky, especially a cool night sky, may be a case in point.

Tracing a series of echoes or emissions from a moving target may, when the received signals are faint, be a needful means not merely of tracking it but of revealing its very presence. A relatively conservative but often effective application of this principle is the cathode ray tube, with the successive projections of signals onto a fluorescent screen persisting long enough to form a visible trace. Nowadays, however, computerised interpretation plays an absolutely crucial part, not least in the operation of phased-array radars. These are the models, often relatively large, that dispense, wholly or partially, with mechanical steering of the beam. Instead, they effect extremely rapid scanning by switching both the pulse emissions and the recaptures thereof as between some thousands of separate sensing elements set, with varying orientations, within the one fixture. Hence these become collectively able to track anything up to 200 or 300 fast-moving targets. They may be also comparatively easy to harden against blast and shrapnel. The main radar of the Boeing E-3A Airborne Warning And Control System (AWACS) is a well-known example of phased array.

At this point, however, discussion of the application of the computer to radar merges into that of its general impact on the military environment. Here the first point to note is that computers divide into two genres, analog and digital. The former measures change on a continuum and the latter does so step-by-step.

Of late, the emphasis in computer advancement has been on the digital mode. For one thing, the analog varieties are usually built for a specific task in the realm of continuous monitoring and control. Your general purpose or 'universal' computer will almost always be digital. What is more, such a machine may effectively do the work an analog model does through the conversion of all inputs into countable quantities. As a rule, too, it requires less in the way of electrical power and complex wiring.

References

1. Ronald J. Pretty (ed.), *Jane's Weapons Systems*, 1979–80 (Jane's Yearbook, London, 1979), p. 590.
2. P. H. Borcherds, 'Focussing a Coherent Beam of Light', *The American Journal of Physics*, vol. 39/6, June (1971), pp. 680–1.
3. The lower atmosphere. The well-defined boundary between it and the stratosphere is known as the tropopause. The height of it varies between about 7,000 m and 20,000 m, depending on latitude and other circumstances. Virtually all cloud is confined to the troposphere (see Appendix B).
4. For a succinct review of this problem in the infra-red and microwave sectors, see Jean Pouquet *Earth Sciences in the Age of the Satellite* (D. Reidel, Dordrecht, 1974), ch. III.
5. Absolute zero is just below minus 273° on the centigrade scale. So minus 273 is taken as 0° on the Kelvin scale. Evidently then, one degree covers the same range on each scale. (See Appendix B.)

APPENDIX B: TERMINOLOGY AND NOTATION

The chief aim here is to facilitate a ready transposition, when reading this text or other literature to do with air power, between the *Système International* (SI) or other metric and the customary Anglo-American or Imperial units of measurements.

(1) The prefix 'kilo' multiplies by a thousand the unit in question.
 The prefix 'mega' does so by a million.
(2) The prefix 'giga' does so by a thousand million (i.e. 1 billion).
 The prefix 'centi' divides by a hundred.
 The prefix 'milli' divides by a thousand.
(2) A foot is c. 0.305 metres.
 A square foot is c. 0.093 square metres.
 A metre is c. 3.281 feet.
 A square metre is c. 10.764 square feet.
(3) A pound is c. 0.454 kilogrammes.
 A kilogramme is c. 2.205 pounds.
(4) In this text, a ton is an American short ton (2,000 lb) rather than a British long ton (2,240 lb).
(5) A metric tonne is 1,000 kg.
(6) As is well-known, the freezing point of water at sea level is 0° on the Centigrade scale, and the boiling point 100°. The corresponding values on the Fahrenheit scale are 32° and 212°. Therefore, 5° Centigrade are equivalent to 9° Fahrenheit.
(7) A state of absolute cold, with all the heat energy drained from everything, is reached a little below $-273°$ on the Celsius Centigrade scale. So $-273°$ is taken as zero for another scale, the Kelvin. Each degree on the Kelvin scale covers the same temperature span as does one on the Centigrade. A calorie is the amount of heat needed to raise the temperature of 1 gramme of water by one such degree.
(8) A watt (the metric unit especially used to measure electrical power) can be defined as 1 joule per second. One joule is equal to 10,000,000 ergs.
 An erg is the amount of work done when a force of 1 dyne moves through 1 centimetre. A force of 1 dyne acting for 1 second on a mass of 1 gramme would accelerate it by a

292 *Appendix B*

velocity of 1 centimetre per second. A Newton is equal to 100,000 dynes.

(9) 1 calorie = 4.19 joules.
1 ton of TNT completely exploding would release the order of a billion calories.

(10) 1 British horsepower = *c.* 745.7 watts.

(11) A bar is a unit of pressure, principally of atmospheric pressure. It is equivalent to 100,000 Pascals. Alternatively, it is 10 Newtons per square centimetre.

(12) For International Standard Atmospheric ratings, the Mean Sea Level average pressure at 45° N is taken as 1,013.2 millibars. It equals *c.* 760 mm, or *c.* 29.90 inches, of mercury.

(13) So a Pascal is a measure of pressure or stress equal to 10 dynes per square centimetre.

(14) As one ascends through the atmosphere into near space, one passes through several ever-present boundary layers. The clearest cut and most important is the Tropopause: the one which separates the lower atmosphere or Troposphere from the Stratosphere. It normally occurs between 10 and 12 km in temperate latitudes but may be below 3 km in the Arctic or above 15 km near the Equator.

The origin of the stratosphere lies in the interaction between free oxygen and the ultra-violet radiation pouring in from the Sun. This radiation leads to the formation of unstable molecules of oxygen (with three atoms rather than the usual two), known as ozone. This, in its turn, blocks the progress of such short-wave radiation to the Earth's surface. So it is absorbed into the stratosphere itself.

Accordingly, that zone is characterised by a massive temperature inversion. Having typically fallen to $-35°$ C at the tropopause, the temperature recovers to 0° or thereabouts as one ascends through the stratosphere. Therefore, the weather patterns of the troposphere can penetrate it only in a limited way. Hence it is virtually cloudless.

The upper surface of the stratosphere is known as the Stratopause. It is encountered round about 50 km up. Above it, the air is so rarified that electrical effects become dominant.

(15) The actual speed being registered in fast flight is normally measured in terms of multiples of the speed of sound in the atmosphere. These are known as Mach numbers after the

Austrian physicist, Ernst Mach (1838–1916). Mach 1.0 (i.e., the speed of sound) is around 760 miles per hour or *c.* 1,115 feet per second just above the surface. At the tropopause, it is usually around 100 mph less.

(16) The acceleration of a body freely falling in the Earth's gravitational field is known as 'g'. It is *c.* 981 cm (or *c.* 32.2 ft) per second per second.

It is used as a standard measure of induced acceleration. Thus, if a force of x pounds acts alone on a mass of one pound it will give it an acceleration of xg.

(17) In Chapter 12, various capital costs are given in terms of the 1975 purchasing power of the United States dollar. In 1975, the US wholesale price index (probably the most relevant indicator of the upward trend in price levels) averaged 175 as against a 1967 baseline of 100. The figure for 1983 was 285.

(18) The pound sterling exchanged for $2.02 at the end of 1975 and $1.43 eight years later. The corresponding figures for the Deutschmark were $0.38 and $0.36; and for the French franc, $0.22 and $0.12.

(19) Warship size is here expressed, as it is very generally, in terms of 'standard displacement': that is to say, weight empty except for fuel and water.

APPENDIX C: SOME CONTEMPORARY MILITARY AIRCRAFT

Strategic Bombers

B-1B

US variable-sweep intercontinental bomber made by Rockwell. In service, 1987. All-up weight, 240 tons. Maximum speed: Mach 1.2. Armament can include up to 22 AGM-86B nuclear-armed cruise missiles or, say, up to 128 Mark 82 500 lb free-falling non-nuclear bombs.

Mirage IV

French two-seat medium-range strategic bomber that entered service 1964. Made by Dassault. All-up weight, 37 tons. Maximum speed: Mach 2.2. Prescribed bombload is 1 ton if nuclear and 8 tons if non-nuclear.

Myasischev-4

Soviet intercontinental bomber, in service 1955. Code-named Bison by NATO. A pure jet of 175 tons, all-up weight. Maximum speed: Mach 0.85. Liberally provided with cannon stations for self-defence. Carries a bombload (nuclear or non-nuclear) of up to 10 tons.

Tupolev-20

Soviet intercontinental turboprop bomber, in service 1955. Code-named Bear by NATO. All-up weight, 160 tons. Maximum speed: Mach 0.80. Liberally provided with cannon stations for self-defence. Carries a bombload of up to 20 tons, including an AS-4 cruise missile.

Tupolev-26

Soviet variable-sweep strategic bomber and maritime strike aircraft. In service, 1975. All-up weight, 130 tons. Maximum speed: Mach 1.9. Internal warload up to 7 tons, including an AS-4 cruise missile.

Tactical Monoplanes

A-10 Thunderbolt 2

Close-support aircraft particularly adapted to the anti-armour role. Made by Fairchild, and in service with USAF since 1975. All-up weight, 24 tons. Maximum speed: Mach 0.6. Armament includes a 30 mm rotary cannon and up to 5 tons of munitions.

Alpha Jet

Light strike-trainer made by Dassault-Breguet and Dornier. Has been entering service with several air forces since 1978. All-up weight, 8 tons. Maximum speed: Mach 0.85. Armament includes one centreline cannon pod (27 or 30 mm) and up to 2.5 tons of munitions.

F-14A Tomcat

United States Navy multi-role fighter made by Grumman. In service since 1973. All-up weight, 37 tons. Maximum speed: Mach 2.4. Armament includes one 20 mm rotary cannon plus maybe six Phoenix missiles plus a couple of Sidewinders.

F-15 Eagle

McDonnell Douglas air superiority fighter in service with the USAF and several other air arms. The two main variants are the F-15A (first flown 1972) and the F-15C (first flown 1979). All-up weight of F-15C, 34 tons. Maximum speed: Mach 2.55. Intercept armament includes a 20 mm rotary cannon and, as a rule, four Sparrow and four Sidewinder missiles. Up to 7.5 tons of ordnance may be carried on external stations.

F-16 Fighting Falcon

Air combat fighter developed by General Dynamics which has entered service with some ten air arms since 1979. All-up weight, 17.5 tons. Maximum speed: Mach 2.0. Intercept armament includes one 20 mm rotary cannon and up to six Sidewiders. Up to 7.5 tons of ordnance can be carried on external stations.

F-18 Hornet

A multi-role fighter produced by McDonnell Douglas primarily for shipboard service in the United States Navy and Marine Corps, which have been receiving it since 1980. But also being acquired by several foreign air forces. All-up weight, 28 tons. Maximum speed: Mach 1.8. Intercept armament includes one 20 mm rotary cannon and two Sparrow plus two Sidewinder missiles. Up to 8.5 tons of other ordnance, carried externally.

F-20 Tigershark

A multi-role fighter independently developed by Northrop, and first flown in 1982. All-up weight, 13 tons. Maximum speed: Mach 2.0. Two 20 mm single cannon borne plus up to 3.5 tons of external ordnance.

F-104 Starfighter

Lockheed fighter-bomber that entered service with various air forces from 1958. All-up weight, 10 tons. Maximum speed: Mach 2.2. Armed with a 20 mm rotary cannon and up to 2 tons of projectiles, perhaps including Sparrow air-to-air missiles.

Hawk

Light strike-trainer manufactured by British Aerospace. In service with the RAF in 1975; and since acquired or ordered by several other air arms. All-up weight, 9 tons. Maximum speed: Mach 0.88. May bear up to 3.5 tons of air-to-air and/or air-to-ground ordnance.

Jaguar

Strike aircraft produced by British Aerospace and Dassault-Breguet. Has entered service with five air forces since 1971. All-up weight, 17 tons. Maximum speed: Mach 1.6. Armament includes two 30 mm cannon and up to 6 tons of munitions.

Mikoyan-Gurevitch MiG-21 BIS

A late variant of the MiG-21 fighter-bomber (code-named Fishbed by NATO), which first entered Soviet air force service in 1960. All-up weight, 10 tons. Maximum speed: Mach 2.1. Armed with a 23 mm twin-barrel cannon and/or four air-to-air missiles or perhaps a ton of air-to-ground ordnance. MiG-21 variants are in service with over 30 countries.

Mikoyan-Gurevitch MiG-23

A Soviet multi-role fighter (code-named Flogger by NATO) which has air superiority and Close Air Support variants. Entered service

1971. All-up weight, 22 tons. Maximum speed: Mach 2.3. Armed with one 23 mm twin cannon and either four air-to-air missiles or up to 5 tons of air-to-ground ordnance. In service with some ten other countries.

Mikoyan-Gurevitch MiG-25

Interceptor and reconnaissance aircraft, code-named Foxbat by NATO. Into serive as an interceptor in 1970. Currently operated by six air forces. All-up weight, 39 tons. Maximum speed: Mach 2.8 or more. May carry four air-to-air missiles.

Mikoyan-Gurevitch MiG-27

A development of the MiG-23 dedicated to the air-to-ground role. Known to NATO as Flogger D or J. D in service with Soviet air force in 1975. All-up weight, 22 tons. Maximum speed: Mach 1.6. Armed with one 23 mm rotary cannon and perhaps 4 tons of ordnance.

Mirage III

Multi-role fighter developed by Dassault-Breguet which became operational with France's *Armée de l'Air* in 1960. Still in service, in one variant or another, with six air forces. All-up weight, 8–15 tons, depending on the variant. Maximum speed: Mach 1.9–2.2. Armed with a pair of 30 mm cannon and up to 4.5 tons of munitions.

Mirage 2000

Latest Dassault-Breguet multi-role fighter. With *Armée de l'Air* since 1982; and on order to several air arms. All-up weight, 17 tons. Maximum speed: Mach 2.3. Armed with pair of 30 mm cannon and either four air-to-air missiles or up to 7 tons of munitions.

F-4E Phantom

One of the major variants of the famous McDonnell Douglas strike fighter, originally designed in response to a United States Navy specification. First Phantom, an F-4B, in service 1960; and followed by F-4E in 1967. All-up weight, 30 tons. Maximum speed: Mach 2.4. Bears one 20 mm cannon and, externally, up to 8 tons of munitions.

Sea Harrier

Version adapted to shipboard use of Harrier: the British Aerospace VTOL strike fighter. First deliveries to Royal Navy in 1979. All-up weight, 13 tons or 10.5 tons for VTOL. Maximum speed: Mach 0.9. Armament includes two 30 mm cannon and up to 2.5 tons of munitions (air-to-air missiles perhaps included).

Sukhoi-24

Soviet variable-sweep deep strike aircraft, code-named Fencer by NATO. Operational since 1974. All-up weight, 44 tons. Maximum speed: Mach 2.3. Armament includes a 23 mm rotary cannon and a 30 mm cannon together with up to 9 tons of munitions.

Tornado F Mk 2

A British air defence variant of the Tornado deep-strike aircraft developed by Panavia: a British, German and Italian consortium. Entered service with the RAF, the Luftwaffe and the Marineflieger in 1982. All-up weight, 26 tons. Maximum speed: Mach 2.2. Armament comprised of a 27 mm cannon and six air-to-air missiles.

Viggen

Interceptor or multi-role warplane developed by SAAB, the first of which appeared in squadron service with the Royal Swedish Air Force in 1971. All-up weight, 25 tons. Maximum speed: Mach 2.1. Armament includes a 30 mm cannon and up to 6.5 tons of air-to-ground ordnance and/or several air-to-air missiles.

Helicopters

Apache

American two-seat attack helicopter made by Hughes. In service from 1984. All-up weight, 9 tons. Maximum speed: Mach 0.25. Armament includes a 30 mm cannon and up to 16 air-to-ground missiles.

Chinook

American transport helicopter made by Boeing Vertol. Entered service first in Vietnam in 1968. All-up weight, 25 tons. Maximum speed: Mach 0.20. Carries up to 44 troops plus external stores.

Kamov-25

Soviet shipborne helicopter, largely dedicated to Anti-Submarine Warfare and code-named Hormone by NATO. All-up weight, 8 tons. Maximum speed: Mach 0.17.

Lynx

British ASW or tactical transport that has been procured by more than 10 armed services since 1976. All-up weight, nearly 5 tons. Maximum speed: Mach 0.30. Carries troops.

Mangusta

Italian light attack helicopter, first deliveries of which due from Agusta in 1986. All-up weight, 4 tons. Maximum speed: Mach 0.22. Armament mainly consists of 8 anti-tank guided missiles.

Mi-24

Soviet assault helicopter in service since 1973, and known to NATO as Hind. The all-up weight of its 'gunship' version, Hind D, is 11 tons; and its maximum speed: Mach 0.23. It can accommodate 8 fully equipped troops; carries a 12.7 mm rotary cannon; and can deliver over a ton of anti-armour missiles and rockets.

Mi-26

Soviet heavy-lift helicopter in service since 1981, and known to NATO as Halo. The heaviest rotorcraft ever flown. All-up weight, 62 tons (22 tons of which may be internal payload). Maximum speed: Mach 0.24. Up to 70 fully equipped troops.

Airborne Early Warning

E-2C Hawkeye

American shipborne AEW aircraft made by Grumman. In service since 1972. All-up weight, 26 tons. Maximum speed: Mach 0.5.

E-3 Sentry

Airborne Warning And Control System aircraft primarily developed by Lockheed. In service since 1976. All-up weight, 162 tons. Maximum speed: Mach 0.75.

Appendix C

Heavy Transport Aircraft

C-130 Hercules

American assault and strategic transport, first version of which was delivered by Lockheed in 1956. All-up weight, 77 tons. Maximum speed: Mach 0.6. May carry up to 128 combat troops. In service with seven air arms in NATO alone.

C-5 Starlifter

Another Lockheed aircraft, quite the most powerful strategic transport in the world. In service since 1969. All-up weight, 380 tons — as much as a third of which may be payload. Maximum speed: Mach 0.8.

NB: This appendix is but an assemblage of the most obvious characteristics of what seem to be most of the more significant aircraft types in military service. In this day and age, however, 'obvious' is by no means equivalent to 'important'. Electronics cannot be encompassed by this sort of review. Likewise the inclusion here of maximum speed serves largely to demonstrate that it is no longer the indicator of relative quality that it once appeared to be.

INDEX

A-4D Skyhawk fighter 184
A-7 Corsair fighter 26, 81, 235
A-10 Thunderbolt fighter 26, 56, 132–3, 187, 228, 295
AGM-86B ALCM 80–1
AH-64 Apache helicopter 171, 173, 298
AIM-26 Falcon missile 219
AS-2 ALCM 4
AS-3 200
AS-5 200, 269
acceleration
 effects of 52–3
Ader, Clément 1, 75
Advanced Technology Bomber (ATB) 71, 210, 228
Advisory Group for Aeronautical Research and Development (AGARD) 263, 274
aeronautical science 71–82
Aérospatiale company 252
Afghanistan 16, 109, 172
Air-Launched Cruise Missiles (ALCMs) 4, 80–2, 200, 232, 269
air power
 ascent of 3–17
 definition of 272
 today 247–55
air-to-air missiles 110, 155–6, 210, 219
 AA-6 missile 110, 219
 AA-7, AA-9 missiles 156
 Falcon missile 219
 Phoenix missile 110, 155, 210
air-to-surface missiles 4, 200, 269
Airborne Early Warning (AEW) aircraft 38, 158–9, 198, 243–4, 299
Airborne Warning and Control Systems (AWACS) 38, 105, 113, 159, 198, 244, 259
aircrew 171, 221, 243
 allocation 24
 effects of acceleration 52–4, 60
 number in cockpit 55–7, 133, 242–3
 stress 52–4
 training 54–5, 242

workload, 24, 61
see also pilots
airfields
 civil 187
 cost of 229
 defence of 181–2
 dispersion from 187–9
 repair of 183–4
 vulnerability of 6, 51–2, 178–89, 221–2
 weapons to disrupt 90–1, 182–3
 World War II 178–9
airlift see transport
Alarm missile 86
Alpha Jet 242, 295
altitude
 and radar 77, 131
 sortie profile 119, 129–30
ammunition 229–30
anti-aircraft artillery 3, 21, 43, 60, 199, 252–5
 SIAM 133–4, 254
Anti-Submarine Warfare (ASW) 91, 194–7
anti-tank weapons 84–8, 100, 243, 253–4
Antonov-2 transport 123
Antonov-22 transport 107–8, 168
Antonov-124 transport 108
Apache helicopter 171, 173, 298
Aquila RPV 232–3, 238
Arab–Israeli wars see Israel
Argentina see Falklands campaign
arrester gear 185
artillery 100–2, 124–7
 aircraft 82–91; accuracy 84–5, 109; anti-airfield 90–1; anti-submarine 91; anti-tank 84–8; cannon 88–9; cost of 230; guidance 85–6; incendiary 87; rockets 86–8
 anti-aircraft 3, 21, 60, 133–4, 199, 252–5
 anti-submarine 91, 194–7
 anti-tank 100, 253–4
 field guns 100, 124, 126–7
 see also missiles
Assad, President of Syria 14–15
attrition rates 26–9, 129–30, 251–2

301

302 Index

'avionics' 78

B-1B bomber 71, 89, 108, 207, 220, 228, 236, 294
B-29 Superfortress 208
B-36, B-50 bombers 208
B-52 stratofortress 24, 64, 66, 81, 208–10, 236
B-70 Valkyrie bomber 208
ballistic missile *see* missiles
Ballistic Missile Defence (BMD) 43, 214–15, 269–70
  see also Strategic Defence Initiative
Beaufre, General 63
Blackbird SR aircraft 73, 143
'Blackjack' bomber 107–8, 215
blitzkrieg 25, 33, 247, 255–63
Bloodhound 2 SAM 219
Boeing 707 159
Boeing 747 81, 232
Boeing Heavy Helicopter (HLH) 169
Bomarc B missile 152
bombers
  cost of 236
  Soviet 107–8, 209–10, 215, 294
  strategic 23, 82, 107–8, 206–13, 215–17, 220–1, 294
  World War II 8–9, 169
  see also aircraft by name
bombing campaigns 7–10, 42, 49–51, 208–9
  blitzkrieg 25, 33, 247, 255–63
  strategy 119, 268
  World War II 24–7, 33, 63–4, 119, 137–8, 247
bombs, aerial 23, 82–7, 90–1, 268
Britain 21, 24, 27, 29, 33, 39, 113, 158, 164–5, 166–7, 181, 187, 198, 203, 222, 274
Brown, Harold 71
Burt, Richard 275

C-5 Galazy transport 81, 108
C-5 Starlifter transport 168–9, 300
C-47 Dakota transport 169
C-130 Hercules transport 168–9, 300
C-141 Starlifter transport 81
CH-47 Chinook helicopter 170, 298
CH-53 Super Stallion helicopter 170
camouflage 146
Caribou transport 170
carriers, aircraft 6, 202–4, 206
  cost of 229, 237–8
  nuclear (CVAN) 229
chaff 63–6, 203, 253
chemical weapons 221–2
China 10–11, 164, 212
Chinook helicopter 170, 298
Circular Error Probability (CEP) 84–5, 208, 268–9
civil aviation 58–60, 67, 107, 187
Close Air Support (CAS) 24–5, 36, 56, 60, 95, 123, 254, 261–2, 273
  aircraft 132–3
  versus interdiction 137, 140
'clutter' 155–6, 198
cockpit design 55–7, 79, 133, 242–3
combat, air-to-air 152–9, 251
  see also fighters
Combat Air Patrols (CAPs) 21, 24, 35, 157, 235
command and control 210–11, 220–1, 258–9
  see also AWACS; FAC
Compass Cope RPV 143–4
computers 22, 76, 110–11
Concorde 107
'Control-Configured Vehicles' (CCVs) 76–7
Corsair fighter 26, 81, 235
costs 21, 227–45
  and design 240–5
  full-life 235–9
  investment and operation 227–35
  multi-role systems 239–40
cruise missiles 4, 28–9, 80, 269
  air-launched 4, 80–2, 232
  at sea 200
  cost of 232, 238
  ground-launched 82, 218, 268–9
  see also by name
Cuba 212–13

Dakota transport 169
Dayan, Moshe 165
Dédalo carrier 204
defence
  air, depth 149–59
  versus offence 95–8
de Guingand, General Sir Freddie 138
Denmark 24, 55
design, aircraft 79, 240–5
  AEW 243–4
  cockpit 55–7, 79, 133, 242–3
  engine 72–3, 107, 243
  vulnerability and 89–90

Index   303

West-East comparison 107–8
wings 3–4, 74–6, 185, 210, 240, 260
deterrence 28, 206–15
doctrines
   air, Soviet 41–6, 248
   air, Western 33–41, 119, 180, 216, 252, 273
   nuclear, NATO 216–18, 266
dual-purpose systems 3, 239–40, 250, 253, 257
Dupuy, Colonel Trevor 137, 140
Durandal airfield weapon 90, 183

E-2C Hawkeye AEW aircraft 243–4, 299
E-3 Sentry AWACS aircraft 38, 58, 113, 159, 228, 244, 259, 289, 299
E-266 Soviet prototype 107
EF-111A, EW F-111 233
Eagle fighters 26, 79, 107–8, 152, 155–6, 234, 241, 250, 257, 259, 273–4, 295
early warning *see* AEW; AWACS
Eisenhower, Dwight D. 164
Elazar, General David 13
electromagnetic environment 63–71, 65, 179–90, 279–89
Electro-Magnetic Pulse (EMP) 220
Electronic Counter-Measures (ECM) 19, 34, 63–6, 70, 81, 125
Electronic Warfare (EW) 34, 56, 63–71, 253
   cost of 233–4, 237
   Soviet Union 113–14, 253
Emergent Technologies (ET) 14, 40
Emerson, Donald E. 22
engine, aircraft, 72–3, 107, 243
Enterprise USN carrier 204
Erickson, Professor John 42
Essex class USN carriers 203
Etendard strike aircraft 221
Ethiopia 16
European Fighter Aircraft (EFA) 56
Exocet missile 201

F-1 Hornet fighter 156
F-4 Phantom 79, 90, 107, 150, 155, 216, 228, 235–7, 248, 297
   reconnaissance RF-4 144, 232–3
F-5A Freedom Fighter 241
F-5E Tiger 59, 241–2
F-14 Tomcat 108, 155–6, 228, 230, 234–5, 295
F-15 Eagle 26, 79, 107–8, 152, 155–6, 234, 241, 257, 259, 173–4, 295
   F-15E dual-role 250, 257
F-16 Fighting Falcon 25, 74, 89, 155, 185, 221, 235–7, 242, 251, 295
F-18 Hornet 57, 235, 295–6
F-20 Tigershark 26, 242, 251, 296
F-86 Sabre 11, 90
F-100 Super Sabre 184
F-104 Starfighter 89, 185, 188, 237, 241, 296
F-111 4, 24, 26, 75, 78, 109, 233
FB-111 236
Fairchild company 56, 132–3
Fairey company 184
Falklands campaign 15, 26–7, 145, 164, 180, 187, 201, 251, 264
fatigue, aircrew 53–4, 257–8
fibre optics 76, 112
fighter aircraft 151–8, 248–51, 295–8
   cost of 227–9, 234–5
   crew 55–7
   design 241–3
   dual-role 239–40, 250, 257
   see also by name
Fighting Falcon 25, 74, 89, 155, 185, 221, 235–7, 242, 251, 295
Fitts, Lt. Col Richard E. 65
Fleet Ballistic Missiles (FBMs) 206, 214
'Flexible response' 218
Follow-On Forces Attack (FOFA) 264
Forrestal USN carriers 202–4, 237–8
Forward Air Controller (FAC) 60, 123, 140, 263
Forward Edge of the Battle Area (FEBA) 22, 263
Forward-Looking Infra-Red (FLIR) 78
Forward Operating Locations (FOLs) 187
France 19, 218
Freedom Fighter 241
fuel, aircraft 71–2, 261
Fuel Air Explosives (FAE) 87, 253
Fuller, Major-General J.F.C. 6, 178

Galaxy transport 81, 108

Garwin, Richard L. 232, 241
Gazelle helicopter 173
General Dynamics company 242
Genie missile 219
geography 102–6, 119–23, 130–1, 272–6
Gepard Flakpanzer gun 124, 126–7
German theatre 96–9, 102–3, 120–2, 264
gliders, assault 166–7
Goalkeeper gun 199
Goering, Field Marshal Hermann 59
Gripen warplane 188–9
Ground Control Approach (GCA) radar 78
Ground-Launched Cruise Missiles (GLCMs) 82, 218, 268–9
  see also VI
guidance systems
  aircraft 77–8
  weapons 85–6, 124–8
  see also radar
guns 100, 124, 126–7
  see also artillery

Hamilcar glider 166–7
Harpoon missile 91, 202, 230
Harrier VTOL fighter 25–7, 187, 198, 204, 221, 266
  Sea Harrier 26–8, 297
Hawk fighter 242, 296
Hawk SAM 127, 269
Hawkeye AEW aircraft 243–4, 299
Heikal, Mohamed 15
Heinkel-111 4
Heisenberg, Werner 281
helicopters 168–76, 298–9
  cost of 234
  crew problems 60, 174
  heavy 169–71
  night flying 60, 174
  Soviet 176, 299
  survivability 171–3
  transport of 168
  US 175–6, 298
Hellfire missile 171
Hercules transport 168–9, 300
Hitler, Adolf 50, 142
Honest John missile 269
Hornet fighter 57, 156, 235, 295–6
Horsa glider 167
Hound Dog ALCM 4
Howse, Major-General H.H. 175
human factor, the 49–61

adaptation to war 57–61, 255–6
airfield as citadel 51–2
seats in cockpit 55–7
stress and fatigue 52–4, 257–8
sustained offensive action 49–51
training and recruitment 54–5, 242
  see also aircrew, pilots
Hurricane fighter 89
Hussein, King of Jordan 16, 24

Identification, Friend or Foe (IFF) 35, 67–9, 154, 263
Ilyushin-2 and -8 57
Ilyushin-76 transport 159
Indo-Pakistan war 27
Integrated Electronic Warfare System (INEWS) 63
Inter-Continental Ballistic Missiles (ICBMs) 109, 206–7, 211–13
interdiction 136–42
  battlefield 39–40, 99, 141–2
  deep 5–6, 42, 99
  history 137–40
  versus CAS 137, 140
  see also FOFA, OMG
International Institute for Strategic Studies (IISS) 19, 273
Ismail, General Ahmed 15
Israel
  field artillery 100
  June War, 1967 12, 24, 27, 29, 167, 179
  Lebanon invasion, 1982 16, 145, 241
  Sinai campaign 11–12, 165
  tactical philosophy 40–1
  technology 14, 40, 106
  War of Attrition 12–13, 40
  Yom Kippur War 13–15, 27, 29, 67, 103, 128, 148–9, 172, 179

JAS 39 Gripen warplane 188–9
Jaguar strike aircraft 140, 187, 221, 296
Japan 6–10, 42, 50, 106, 201, 203
Jet-Assisted Take-Off (JATO) 184–5
June War 12, 24, 27, 29, 167, 179
Junkers
  Junkers-87 27
  Junkers-88 239
  Junkers-287 75

Kamov-25 helicopter 299
Kennedy administration, 216–17

Khrushchev, N. S. 219–20
Korean peninsula 103, 121, 163, 216, 264
Korean war 3, 10–11, 15, 103, 138–9, 165, 180, 208

Lancaster bomber 67
Lance missile 100, 232
Lanchester, F. W. 5, 19–20, 240, 249, 275
  his square law 11, 19–20, 23, 27, 101, 172, 251
land warfare 94–103
  air power and 98–100, 256–7
  artillery 100–2, 124, 126–7
  geostrategy 102–3
  'Liddell Hart fallacy' 95–8
  movement and fire 94–5
landing of aircraft 184–6
  see also airfields; VTOL
lasers 57, 78, 281
  as weapons 134–5
Lebanon 16, 145, 241
Le May, General Curtis 209
Liddell Hart, Sir Basil 95–6, 98, 157, 255
  fallacy 95–8, 146
Lockheed company 241
  see also F-104; SR-71; U-2
Long-Range Combat Aircraft 207
Long-Range Maritime Patrol (LRMP) aircraft 21, 143, 197
'look-down, shoot-down' 155–6, 198
loss rates 26–9, 129–30, 251–2
Lynx helicopter 168, 299
Lysander army observation aircraft 123

M-113 armoured personnel carrier 231
Mi-24 helicopter 299
Mi-26 helicopter 169, 299
MiG-15 fighter 11, 90, 180
MiG-21 fighter-bomber 12, 59, 90, 153–4, 184–5, 221, 229, 241, 296
MiG-23 fighter 109, 156, 221, 229, 296–7
MiG-25 fighter 108, 110, 143, 154, 156, 185, 219, 297
MiG-27 fighter 156, 221, 297
MacArthur, General Douglas 10
McDonnell-Douglas company 241
Mach, Ernst 292–3

McNamara, Robert 21–3, 175
Magister strike-trainer 27, 29, 243
Manchuria 42, 142, 180
Mangusta helicopter 299
mass destruction see nuclear war
materials in aircraft manufacture 73–4, 107–8
Matra company 90
Maverick missile 86, 129, 230
Messerschmitt company 50, 89, 103, 166
Meteor fighter 248
Middle East 11–17, 24, 27, 40–1, 42, 66–7, 103, 106, 107, 113, 121, 123, 128–9, 141, 145, 146, 148–9, 164–5, 172, 179, 257, 264, 266
Midgetman missile programme 214
Mikoyan-Gurevitch, see MiG-15; MiG-21; MiG-23; MiG-25; MiG-27
mines 87, 193–4
Minuteman ICBM 212–13
Mirage III TR/fighter 12, 142, 185, 221, 297
Mirage IV bomber 215, 220–1, 294
Mirage 2000 fighter 155–6, 297
Mirage F-1 156
missiles
  air-to-air 110, 155–6, 210, 219
  air-to-surface 4, 200, 269
  ALCMs 4, 80–2, 232
  anti-ballistic 43, 269, 270
  anti-satellite 274
  cost of 230, 238
  cruise 4, 28–9, 80–1, 200, 232, 238
  dual purpose 3
  FBMs 206, 214
  GLCMs 82, 218, 268–9
  ICBMs 109, 206–7, 211–13
  IRBMs 217
  SAMs 3, 14–15, 19, 39, 124, 127–8, 133–4, 149–53, 200, 219, 252, 269–70
  SIAMs 133–4, 254
  SLBMs 206–7, 211, 213
  Soviet 109–10, 156, 213, 219
  surface-to-surface 100–1
Mobile Army Surgical Hospital (MASH) units 165
mobility, tactical 163–76
  equipment transport 167–9
  helicopters 169–76

306 *Index*

paratroops 163–6
planes 166–9
Mosquito multi-role aircraft 89, 143, 239
Multiple Independently-targetable Re-entry Vehicle (MIRV) 231
Multiple-Launch Rocket System (MLRS) 101
'multi-role' aircraft 239–40, 250
Mussolini, Benito 7
Mutual Assured Destruction (MAD) 212, 215–16
Myasischev-4 bomber 294
Myrdal, Alva 39

'Nap-of-the-Earth' flying 172, 174
Napoleon Buonaparte 44
Nasser, Gamal Abdel 13, 15, 24
NAVSTAR 77
navigation 38, 77–9, 269
navigator 54–5
Nike Hercules SAM 3, 127, 219
Nimrod AEW aircraft 198, 259
North Atlantic Treaty Organisation (NATO)
    air warfare doctrine 33–41, 119, 180, 216, 252, 273
    crew-plane ratios 57
    IFF 67–8
    manning levels 273
    nuclear doctrine 216–18, 266
    sortie rates 25–6
    strategy, quest for 263–6
Northrop corporation 26, 241–2, 251
Norway 103
nuclear-driven aircraft 71–2
nuclear war
    attrition 28
    deterrence 206–15
    escalation 215–16, 265
    'flexible response' 218
    MAD 212, 215–16
    NATO doctrine 216–18, 266
    theatre confrontation 215–22
    threshold 265

Odell, Peter 236–7
Operational Manoeuvre Group (OMG) 44, 46, 265
Operational Research and Analysis 20–1
Optica TR aircraft 143
ordnance *see* artillery; mines; missiles; torpedoes
Organisation for Economic Co-operation and Development (OECD) 111
Orion LRMP aircraft 81
Over-the-Horizon (OTH) radar 158, 284

P-3C Orion LRMP aircraft 81
P-47 Thunderbolt fighter 227–8
paratroops 7–8, 163–6
particle-beam weapons 214–15
Patriot SAM 127, 219
Pegase winged dispenser 129
perception, pilot 52, 69–70, 147–8, 153–4
    and weapon aiming 125–6, 131–2
Permissive Action Link (PAL) 221
Pershing 1 missile 238, 270
Pershing 2 missile 100, 109, 182, 218, 221, 232
Phantom fighters 79, 90, 107, 144, 150, 155, 216, 228, 232–7, 248, 297
pilots
    crew plane ratios 57–8
    electronics and 79–80
    fatigue 53–4, 257–8
    helicopter 60, 174
    perception 52, 69–70, 131–2, 147–8, 153–4
    recruitment 54–5
    weapon aiming 79–80, 125–6, 131–2
Poland 186
Polaris SLBM 206, 207, 212–13
Power, General Thomas S. 208
Precision-Guided Munitions (PGM) 86–7, 97, 105
Protivo Vozdushnaya Oborona Strany (PVO Strany) 41, 45, 248
Pucara light strike aircraft 187, 251

RBS-70 SAM 128
RF-4 Phantom reconnaissance aircraft 144, 232–3
radar
    airborne 77, 131, 143, 154–5
    and IFF 67–9
    at sea 203
    'clutter' 155–6, 198
    counter-measures 63–71
    GCA 79
    OTH 158, 284
    SLAR 143

Soviet Union 109–10
  terrain-avoidance 77, 131
  weapon guidance 86, 124–8
  see also electromagnetic environment; EW
Rampart missile 183
Rand corporation 22
Reagan, Ronald 214
reconnaissance 142–9
  accuracy of information 147–9
  aircraft 142–4
  beyond FEBA 98–9, 148
  NATO doctrine 34, 36
  RPVs 34, 144, 147–8, 270–1
  satellite 144–5
recruitment, pilot 54–5
Remotely Piloted Vehicles (RPVs) 34, 110, 144, 147–8, 232–3, 238, 253, 270–1
Richardson, Lewis F. 17
Rogers, General Bernard 40
Rommel, Field Marshal Erwin 138, 255
Royal Air Force (RAF) 5, 216–17, 273–4
  airfields 181
  altitude 129–30
  budget 238
  manning levels 57–8
  sortie rates 25
  torpedo-bomber squadrons 199
  training 54
  World War II 9, 24, 39, 50, 63–4, 89
Royal Navy 6, 26–7, 206–7, 274
Rumsfeld, Donald 209

S-3 Viking ASW aircraft 81
SR-71 Blackbird 73, 143
SS-18, SS-19, SS-21, SS-23 Soviet missiles 109
SS-20 Soviet missile 109, 213
Sabre fighter 11, 90
Sadat, Anwar 14–15
satellites 60, 77, 105, 144–5
  anti-satellite missiles 274
Scharnhorst battle-cruiser 203
Scorpion tank 230–1
sea, air power at 193–204, 266–7
  AEW 198
  aircraft carriers 6, 202–6, 229, 237–8
  LRMP 21, 143, 197
  missiles 200–1

Soviet Union 44
  submarines and ASW 91, 193–7
  World War II 6–7, 199, 202–3, 213–14
  see also SLBMs
Sea Harrier VTOL fighter 26–7, 298
'second echelon' 40, 99, 265
  see Operational Manoeuvre Group
Self-Initiating Anti-aircraft Missiles (SIAMs) 133–4, 254
Sentry AWACS 38, 58, 113, 159, 228, 244, 259, 289, 299
Shahine SAM 128
Sharon, General Ariel 14
Shazli, Lt-General Saad el 12, 149
Short Take-Off and Landing (STOL) aircraft 184, 259–60
Shrike missile 86
Sideways-Looking Airborne Radar (SLAR) 143
Sikorsky corporation 174
Skybolt missile 217
'Skyhawk' fighter 184
Skyraider AEW aircraft 158
Skysnare balloon 182
Slessor, Sir John 136
Smuts, Field Marshal Jan 5
Sokolovsky, Marshal V.D. 18
sonar 194–7
sorties
  attrition/loss rates 26–9, 129–30
  rates of 23–6
  strategy 119, 129–30
Soviet Union
  Afghanistan invasion 16, 109, 172
  air warfare doctrine 18, 20, 41–6, 248
  bombers 107–8, 209–10, 215, 294
  EW 113–14, 253
  fighters 11–12, 90, 108–10, 156, 221, 296–7
  helicopters 176, 299
  missiles 109–10, 156, 213, 219
  technology, compared with USA 106–10, 113–14
  World War II 3, 8, 42, 44–5, 57, 142, 157, 178–9
space 214–15, 273–4
  satellites 60, 77, 105, 144–5
  weapons 214–15
Spica FPB 202
Stalin, J. V. 275
stand-off delivery of weapons 91, 129, 215, 241, 267

Index

stand-off (stand-back) situation 99–100, 272
Starfighter 89, 185, 188, 237, 241, 296
Starlifter transports 81, 168–9, 300
Stealth technology 70–1, 210, 222
Sting Ray torpedo 91
Strategic Air Command (SAC), USAF 208–9, 212–13
Strategic Defense Initiative 45, 214–15
strategy, air 119, 129–30, 263–8
see also doctrines
stress, aircrew 52–4, 257–8
strike trainers 242–3
Student, Major-General Karl 7
Submarine-Launched Ballistic Missiles (SLBMs) 206–7, 211, 213
submarines and ASW 91, 193–7, 206–7
Suez campaign 29, 165, 179, 208
Sukhoi fighters
  Su-7 185, 221, 229
  Su-17 109, 221, 243
  Su-19 185
  Su-20 229
  Su-24 221, 298
  Su-27 108
Super Sabre fighter 184
Super Stallion helicopter 170
Superfortress bomber 208
Surface-to-Air Missiles (SAMs) 3, 19, 39, 124, 127–8, 219, 269–70
  and aircraft 149–53, 252
  at sea 134, 200
  Israeli-Arab Wars 14–15, 128
  nuclear warheads 219
  see also by name
surface-to-surface missiles 100–1, 109, 268
Sweden 128, 158, 202
  Royal Swedish Air Force (RSAF) 39, 186, 188–9

Talos SAM 200
tanks 100
  and fallout 221
  anti-tank weapons 84–8, 100, 243, 253–4
  cost of 228, 230–1
  transport of 167–8
technology 63–91
  aeronautical science 71–82

EW 63–71
extended S-curve 111–12
gaps in 105–14
global diffusion 105–6
ordnance 82–91
West-East comparison 106–10
Tedder, Lord Arthur 153
theatre confrontation 215–22
Thunderbolt fighters 26, 56, 132–3, 187, 227–8, 295
Tiger 2 fighter 59, 241–2
Tigershark fighter 26, 242, 251, 296
titanium 73, 107–8
Tito, Marshal 7
Tomahawk ALCM 80–2, 232
Tomcat fighter 108, 155–6, 228, 230, 234–5, 295
Tornado fighter 69, 78, 91, 129, 140, 156, 185, 240, 298
torpedoes 91, 194, 199
'toss-bombing' 199–200
training, aircrew 54–5, 242
transport by air 166–76
  equipment 167–9
  helicopters for 169–76
  NATO doctrine 37
  paratroops 7–8, 163–6
  planes 166–9, 300
  World War II 8, 123
Trenchard, Lord 49
Trident missile system 231
Tukhachevsky, Marshal 275
Tupolev-16 bomber 200
Tupolev-20 bomber 200, 294
Tupolev-22 bomber 107, 209–10
Tupolev-26 bomber 294
Tupolev-114 transport 159
Tupolev-126 AWACS 159, 198
Tupolev-144 transport 107

U-2 SR aircraft 143–4
Union of Soviet Socialist Republics (USSR) *see* Soviet Union
United Kingdom *see* Britain, Royal Air Force, Royal Navy
United States of America (USA)
  Air Force (USAF) 10, 29, 33, 38, 55, 59, 68, 208, 210, 215–16, 274–5
  air warfare doctrine 33–41, 119, 130
  Army 33, 175–6
  Army Air Force (USAAF) 9–10
  Kennedy era nuclear strategy 216–17

Navy (USN) 6–7, 193–4, 201–4, 206
  technology compared with Soviet Union 106–10, 113–14

V bombers 180, 208, 210, 216–17
V1 missile 4, 28–9, 51
Valkyrie bomber 208
Vertical Take-Off and Landing (VTOL) aircraft 25, 43, 184, 204, 260–2, 266, 274
Vietnam war 3, 15, 22, 139, 247
  air combat 59, 157, 241
  bombing 15, 208–9
  EW 66–7
  helicopters 165, 172, 276
  sortie and loss rates 24, 29, 276
Viggen fighter 39, 298
Viking ASW aircraft 81
Vulcan gun 126, 230

war
  adaptation to 57–61, 255–6
  developments 246–76; air power 247–9; blitzkrieg 255–63; NATO strategy 263–6; warplanes 250–5
  geography 17, 97–8, 102–3, 136, 119–23, 272–6
  history *see wars etc. by name*
  nuclear 206–22
  sea, air cover from 266–7
  systems, mix of 267–72
  weather 15, 78–9, 119–21, 166, 286
warning systems *see* AEW; AWACS
warplanes 17–18, 250–5

see also bombers; fighters; *etc.*
Wasp missile 243
weapons *see* artillery; missiles
weather 52, 78, 120–1, 145, 173–4, 257
Wells, H. G. 249
Whitcomb, Richard 74–5
Wilde Sau tactic 64
wings, aircraft 3–4, 74–6, 185, 210, 240, 260
World War II
  airfields 178–9
  Atlantic 199, 213–14
  Battle of Britain 24
  bombers 8–9, 169
  bombing campaigns 8–10, 24–5, 27, 49–51, 63–4, 119, 137–8, 247
  Eastern Front 3, 8, 44
  fighters 89, 157
  interdiction 136–8
  loss rates 27
  North Africa 138, 146
  Operational Analysis 20–1
  Pacific 6–8, 42, 50, 201–3
  paratroops 7–8, 164, 166
  sortie rates 24–5
  Soviet Union 3, 8, 42, 44–5, 57, 142
  V1 missiles 4, 28–9, 51
  Western Front 5

Yakolev-14 glider 167
Yom Kippur War 13–15, 27, 29, 67, 103, 128, 148–9, 172, 179

ZPU-4 multiple machine-gun 126